Online Learning for STEM Subjects

The Global Collaboration initiatives related in this book are examples of how educators have experimented with different mechanisms to provide science, technology, engineering and mathematics (STEM) education programmes through information and communication technologies. In many cases, these programmes have looked at the allied personal communication and collaboration skills that students of these subjects also need: the so-called STEM+ curriculum.

In particular, these approaches to STEM+ provision show how the internationalization of education can be made more effective and accessible through the exploitation of collaborative technologies and non-traditional pedagogies. The approaches address the following themes:

- technologies for distance learning and collaboration
- pedagogies for online learning
- remote communication and collaboration.

An international perspective is made possible within the book through the inclusion of authors from North America, Europe and Asia. These authors present case studies from technology-enhanced learning projects over the past six years.

Mark Childs is a Senior Lecturer in Technology Enhanced Learning at Oxford Brookes University, UK, supporting staff in developing their digital capabilities. Mark has been working in higher education since 1995 as a researcher in the use of technology for learning and teaching. Mark's main research area has been the study of learners' experiences of synchronous technologies in higher education and the role of presence and embodiment in supporting learning in these. His PhD in education was completed in 2010.

Robby Soetanto is Senior Lecturer in Construction Management at Loughborough University, UK. Prior to this, he held post-doctoral research posts at Wolverhampton and Loughborough Universities and lectured at Coventry University. His research has been funded by government bodies (EPSRC, EU, HEA, British Council) and private companies (Lloyds TSB and Hewlett Packard). He led Learning to Create a Better Built Environment and BIM-Hub projects, which focus on developing virtual collaboration skills of built environment students and provide the basis of the first part of this book.

Online Learning for STEM Subjects

International Examples of Technologies and Pedagogies in Use

Edited by Mark Childs and Robby Soetanto

Routledge
Taylor & Francis Group

LONDON AND NEW YORK

First published 2017 by Routledge

2 Park Square, Milton Park, Abingdon, Oxon, OX14 4RN
605 Third Avenue, New York, NY 10017

Routledge is an imprint of the Taylor & Francis Group, an informa business

First issued in paperback 2020

British Library Cataloguing-in-Publication Data
A catalogue record for this book is available from the British Library

Library of Congress Cataloging-in-Publication Data
A catalog record for this book has been requested

ISBN: 978-1-138-93444-3 (hbk)
ISBN: 978-0-367-73647-7 (pbk)

Typeset in Times New Roman
by Apex CoVantage, LLC

Contents

PART II
Complementary case studies

Figures and tables

Figures

Tables

Editors' introduction

Online collaboration is no longer an emerging issue of the 21st-century workplace but is an ever-present element of many working lives. This is particularly true of large-scale science, design and engineering projects, which require the combined expertise of many different professions to create the necessary end-products, whether those are games, airplanes or buildings. The time taken to travel and the environmental impact of travelling preclude those practitioners from meeting face to face, and each time there is a global panic about terrorism or pandemics, there is a noticeable reduction in travel plans and a consequent increase in online collaboration.

The building industry in particular has a dependence on the sharing of complex and contingent information online. Building information modelling (referred to as BIM) is a predominantly transactional online environment in that it is concerned chiefly with passing on information, documents and files and enabling them to be tagged and appended, and flagging when updates are created or required. BIM has not been, up to now, an environment commonly built around enabling the synchronous co-creation of content.

Indeed, the synchronous co-creation of content has been identified as a particularly difficult practice to translate to an online environment. The design and construction of physical assets is a multidisciplinary activity, which requires contributions from parties who may have different interests and pre-conceived ideas of the project. Although diverse disciplines involved in construction can introduce many innovative ideas for the benefit of a project (Horwitz and Horwitz, 2007), it is recognized that many causes of poor performance emanate from communication problems between parties during the course of a project (Dainty *et al.*, 2006). Often, the problems remain hidden until the construction plan and design are implemented on site. One of the notable influences is the educational background and training of individual parties in their earlier years of engagement with the built environment sectors.

Part of this complexity arises from the so-called "messy talk" phase of co-creation (discussed in more detail later), in which collaborators share their ideas and gradually and spontaneously formulate a shared mutual conception of what is required. While the dead ends, revelations and misunderstandings that occur when collaborators first develop a shared understanding of a project are not unique to STEM subjects, it is perhaps of greater importance to overcome these more quickly when a large industrial project is at stake. The need to identify quickly lapses in shared understanding, feed that misunderstanding back and obtain a better understanding requires synchronous working. The building of trust and the social contractual obligations that underpin the development of trust require an environment that supports sociability and co-presence. Creating this sense of togetherness in a shared space and a wide free-ranging conversation is commonly created in a face-to-face environment. Although advanced developments in information and communication technologies (ICT) have made possible real-time, distanced communication between parties in different locations (Gaudes *et al.*, 2007), technology still falls short in replicating the experience of meeting face to face when at a distance.

Additionally, the "interface" problems which may have existed between these parties could be further exacerbated by the need to communicate over distance within a time constraint in an increasingly interconnected and globalised construction sector. Often, individuals have to work collaboratively with people in different parts of the world whom they have not even met before, let alone worked with together. Many effective practices that are applicable to traditional co-located teams may no longer be relevant and require a thorough examination.

Given the importance of negotiating a common understanding of problems and given the essential nature of building a working rapport, the move to working online presents an enormous and self-evident challenge to the future of collaborations. Further advancement of the BIM way of working (i.e. beyond level 2) will see the online environment moving from simply moving, tagging and curating information to one of co-creation of new content in real time, so the need to develop the skills to conduct online collaboration is even more pressing.

Developing a better communication practice in the industry requires fundamental rethinking of the education content and process of the future built environment professionals. The real integration in the workplace would be a difficult task, but educational institutions can contribute by introducing and incorporating aspects of multidisciplinary working into the curricula in the early years of professional education. Multidisciplinary working presents a significant conceptual challenge for the students, as this would require a comprehensive understanding of the interests and orientation of the other subject disciplines and fit them in the 'jigsaw' of knowledge to produce the constructed facility. This understanding may get better as individuals obtain more experience from their exposure in workplace practice. Further, there are attitudinal requirements that will facilitate successful multidisciplinary working, for example, a willingness to accept other ideas, level of trust, a preference to working in teams, the ease of establishing relationships with others in the team, which are very much related to the culture at functional, organizational and national levels. These all should be better acquired through experiential learning rather than infused through the process of knowledge transmission during traditional lecture sessions. This, coupled with increasing competition in the higher education sector, has promoted the introduction of new pedagogical approaches to teaching and learning.

However, "training" in using technology is still based almost entirely at the level of which button to press to make it work, this despite the awareness of professional bodies for the need for soft skills in STEM education. As we will see in the chapters in this book, skills in teamworking and communication do not automatically translate to skills in teamworking and communication *online*. Most undergraduates are younger than the World Wide Web, and yet the skills acquired through a familiarity with social media and the "prosumer" culture do not prepare this generation of learners with the required skill set for working in an online environment. Prensky introduced the idea of the "digital native" in 2001 (Prensky, 2001) and rescinded it in 2008 (Prensky, 2009), yet the assumption that students are automatically digitally literate due simply to the timing of their birth persists.

Therefore, although the aim in presenting the work in this book has a variety of intentions, the chief of these is to exemplify the need for a concerted effort in preparing students (and indeed those who are already in work) with the skills and experience required for effectively participating in this new working environment. Acquiring technological expertise extends beyond how to use technology to *how to make use of* technology; that the soft skills recognised in STEM education fall short of preparing students if they do not also prepare them for employment in distributed teams and that these are not the same skills as those required for face-to-face working.

If we require our students to be fully literate, fully participative and fully prepared, then acquiring skills for online collaboration is imperative.

About this book

The book presented here is a hybrid book. Our experience is that the format of such a book gives the best possible coverage of a subject. The first part of the book is authored by a group of people who worked on a linked pair of projects, introduced in Chapter 1 by Robby Soetanto and others. These five linked chapters give an opportunity to explore a single case in depth, to identify the learning from the projects and provide the students' voices in reflecting on their experience. These two projects, called Learning to Create a Better Built Environment and BIM-Hub, ran at Coventry University (UK) and Ryerson University (Canada) with Loughborough University (UK) joining the consortium for the second project. The two projects formed two cycles of planning, implementation and evaluation and therefore formed two iterations of a single action research study and are described in depth in Chapter 1 by Robby Soetanto and others. Chapters 2 and 3, by Mark Childs, each discuss the findings of one of these cycles, indicating how the design and experience of the activities developed (or did not develop) over the iterations. These chapters focus on the qualitative experience as voiced by the students and explore alternative ways of framing and structuring an analysis of this experience. Chapter 4, also by Robby Soetanto and others, examines the student experience but takes a more focused and quantitative view on one aspect of this experience, that of the impact of the project on student views on their employability attributes. Finally, Chapter 5 by Harry Tolley and Helen Mackenzie reflects on the evaluation process of the projects. This form of meta-evaluation rarely takes place in projects, and yet evaluating the effectiveness of the evaluation is essential if we are to learn to improve the practice of evaluation.

The second part of the book complements this in-depth look at a single case study by including authors who represent the wider general field of online collaboration for STEM education. Four chapters were included, and these were selected to add to the conversation of the book in four distinct capacities.

Any discussion on online learning requires the inclusion of MOOCs. MOOCs, or massive online open courses, are less than a decade old at the time of writing, and their potential is still being explored. They offer the possibility of large-scale, affordable education and have been identified as a flexible mode of delivering built environment education (BIM2050 group, 2014), and yet issues such as routes to certification and sustainable business models are still to be adequately addressed. Neil Smith, Helen Caldwell and Mike Richards report on two case studies that have informed this debate and perhaps may point the way to identifying how MOOCs may fulfil the wide-scale need for STEM education; in keeping with the theme of the book, the key is in finding effective ways to encourage sustained online collaboration.

Peng Li, in our second chapter of cases extending our view of online collaboration, presents a study on how cloud computing can be a useful platform to enable this collaboration to take place. Li reports on how virtual computing labs can be implemented for both face-to-face and distanced collaborations between students.

The third chapter from our discussion of wider contexts is provided by Giovanni Migliaccio, Ken-Yu Lin and Carrie Dossick. In this chapter, they report on two additional projects under the Learning to Create a Better Built Environment banner. The first of these extends Li Peng's discussion of the use of cloud computing that took place in Kenya and had aims of promoting global collaboration. The second discusses a simulation in the form of a game to educate built environment students in safety inspection.

In the final chapter of our case studies extending the discussion of STEM education, Farzad Pour Rahimian, Jack Steven Goulding and Tomasz Arciszewski also discuss extending the use of gaming technologies to make them applicable to education. In their project, a virtual-reality recreation of a construction site was used to develop a pedagogical approach called *successful intelligence*, as in the Migliaccio and colleagues and Soetanto and colleagues chapters, discussing specifically how these approaches support employability in STEM subjects.

In the remainder of this introduction we will provide a basic grounding in many of the areas explored in this book for readers who may be unfamiliar with some of the concepts explored throughout these chapters.

Educational practices in STEM subjects

New approaches to STEM+ education

Skills shortage has represented an enduring problem in the engineering and construction sectors for a number of years (Forde and MacKenzie, 2004), an issue that potentially hampers the economic growth and competitiveness of industry. Amongst the factors that may contribute to this issue, some publications attribute this to the failure of the education sector to attract a young workforce to choose their career in engineering and construction sectors and hence to study in the relevant science, technology, engineering and math (STEM) subjects (e.g. Dainty and Edwards, 2003). Coupled with increasing competition within the education sector, this issue has encouraged some universities (such as Aalborg in Denmark and Coventry University in the UK) to rethink and promote the introduction of a new pedagogical approach to teaching and learning, called activity-led learning (ALL). ALL is underpinned by principles of problem-based learning, inquiry-based learning, project-based learning, in that the learning is located around student activities; the students learn more effectively from the activities they undertake and experience first-hand rather than from listening to traditional lectures. In ALL, an activity is a problem, project, scenario, case-study, research question or similar in a classroom, work-based, laboratory-based or other appropriate setting and for which a range of solutions or responses are appropriate (Wilson-Medhurst *et al.*, 2008). As most ALL applications involve some form of group activities, the learners gain not only technical knowledge but learn key employability skills such as communication, collaboration and teamworking skills. A UK government commissioned review on the skills for creating sustainable communities highlighted the importance of communication and professional skills for built environment graduates (civil and building engineering), which presently require further development (Egan, 2004). This is in line with the required skills for 21st century in STEM subjects (STEM+). ALL is aimed to harness both hard technical and "soft" professional skills for future employment while at the same time enhancing student experience and motivation through engaging activities in the learning process.

Learning by design

In aiming for a shift from a teacher-centred to a student-centred approach, learning by design has grown alongside other approaches such as problem-based learning, activity-led learning and so on. Learning by design has been described as a constructivist approach in that "activities involving designing, making or programming – in short designing – provide a rich context for learning" (Kafai and Resnick, 2011; 4), that is the process of constructing an object, whether this is an actual physical object in a classroom, laboratory or workshop or a digital artefact on a computer screen, engages students with a utilitarian relationship to their own learning. In learning by design, students seek out the information they require to fulfil the specific demands of the construction of the artefact at that moment and add this information to their own knowledge in the specific context of that creation process and so with an immediate awareness of how that knowledge can link to other pieces of knowledge. They are therefore constructing meaning and understanding of the world around them in parallel with constructing the artefact. The metaphor with building is self-evident; their body of knowledge is a construction, engineered by themselves, though more effective if scaffolded by the teacher.

The range of other learning experiences resulting from learning by design is an extensive one; in fact, the revised Bloom's taxonomy (Krathwohl, 2002; 213) places Creating (including designing, constructing, planning, producing, inventing, devising and making) at the highest level, standing above the original Bloom's category of evaluating. The implication is that to achieve creation of an artefact often requires each step of the process to encompass those activities further down the hierarchy of activities; in order from lowest- to highest-order thinking these are Remembering, Understanding, Applying, Analysing and Evaluating. So, for example, in the lead of the cases represented in this book, in order to create a building design students must remember building regulations, understand how these relate to other principles such as materials strength, apply these to a particular requirement of the client, analyse the competing potential designs and evaluate the most effective way forward. Setting our students the task of creating an artefact and then allowing them to choose the learning they require to meet that goal not only provides them with an authentic learning experience, in that this is how they will acquire knowledge once they leave formal education, but also very likely encompasses an entire range of intermediary types of learning activity.

Collaborative learning

Another component of learning by design is that design is a collaborative activity; the stages of applying, analysing and evaluating that students go through in order to create (according to the revised Bloom's taxonomy) are steps that students will work through together if given the opportunity and direction (Gokhale, 1995). Collaborative learning requires social negotiation (Driscoll, 2005; 397) and enables groups of students to learn by co-constructing and sharing their joint understanding of a discipline, a view of learning described as social constructivism (Conole *et al.*, 2005; 11). Creating learning opportunities through setting collaborative tasks therefore creates "a situation in which particular forms of interaction among people are expected to occur, which would trigger learning mechanisms" (Dillenbourg, 1999; 5), including learning from each other.

Collaborations share a number of common factors, identified by Dillenbourg (1995; 2–7). These are:

- The collaborators develop a shared culture, one in which they have a common understanding of words, goals, approaches.
- The collaborators develop a distributed cognition, through a shared set of resources, through traces of interactions such as through social media and a collective memory of a group experience, a learning approach often referred to as connectivism (Cormier, 2008).
- A shared social contract arises, both between themselves and (because in a learning context, a task has usually been set by a teacher) between the group and the teacher.
- A shared task, situation and constraints applies; it is usually these that the teacher has the most control over, and the design of these is therefore the area

in which the teacher can best contribute to the success of the collaboration. The role of the teacher in collaborative learning is discussed in what follows.
- Related to the task set them the group will also develop their own shared goals and, through negotiation, a mutual understanding of these shared goals.

However, in parallel to these commonalities between collaborations, there is also a lack of commonality concerning what the basic nature of the collaborative learning actually is. A common definition collaborative learning as "a situation in which two or more people learn something together" and yet the words "more", "something" and "together" are highly problematic (Dillenbourg, 1999; 1–5).

The variety of scales at which collaborative learning can take place begins with two people, or even a person and an AI agent (Childs *et al.*, 2013), up to the thousands of people who interact in a MOOC, although as can be seen from the chapter by Neil Smith and colleagues, collaboration does not automatically occur, and when it does, it can be on smaller scales than the entire cohort of learners. The nature of the collaboration is vastly different depending on these scales, and as can be seen in Mark Childs's chapters, even a move from four to six members of a collaboration can have a large impact on the nature of the collaboration.

The nature of what is being learnt also varies. The "something" in the definition can refer to a single task, a module within a course or an entire degree. Within the cases represented here we have exclusively explored a single module or courses of module length. However, these too can be very different. The modules described within the book range from those of the main case discussed in the first half of the book, which were located around two different projects but within a traditional undergraduate degree, to an open course in which anyone was able to participate. In the chapter by Peng Li, this was a task that ran across several modules within a degree programme.

Possibly the most problematic of these three words is "together", as although many may also define being together as face to face or computer mediated and synchronous or asynchronous (Dillenbourg, 1999; 2), for many people "together" still presupposes face-to-face interaction. In this book we have chosen solely to focus on "together" being chiefly online. The problematic nature of "together-ness" online is explored in the following section.

As stated, setting up the initial conditions for interaction, by the design of the shared task, by the facilitation of the situation and by limiting the constraints on the collaboration, are the areas in which the teacher can have most influence over the collaboration. The success or otherwise of the various approaches taken by the authors is the central theme tying together the different chapters. Dillenbourg (1999; 5–6) identifies the ways in which a teacher can help support the collaborative learning as:

- Set up the initial conditions. This can be in terms of allocating who should work in the groups, what technologies should be used, what tasks should be set and should these be divided into subtasks for the student? What training do the students need? What assessments should be used?

- Specify the "collaboration" contract. To what extent should the nature of the collaboration be determined by the teacher, for example, allocating roles for the students? (The difficulties students can face in allocating roles can undermine the effectiveness of a collaboration.)
- Providing rules for interaction. Face-to-face interactions can prove difficult for collaborators; having a chance to present information in the face of dominant members of a group, power dynamics between individuals, accessibility issues and so on. In general, however, we are accustomed to these in face-to-face situations due to the face-to-face environment being a long-standing one (i.e. the situations that arise in face-to-face collaborations are those we contend with throughout our lives and also have been present throughout human history), and so we do not place the difficulties as being contingent on, and therefore because of, the medium of interaction. In the move to an online environment, these difficulties are replaced with a different set of constraints on interaction, but as technology is a relatively new medium for communication, technology-mediated communication is seen as more difficult than face to face, and the problems we face in communicating are seen to be due to the environment. Furthermore, although students often have experience of collaboration offline, in schools or in earlier years of higher education, these tend to be with immediate peers and friends, which are then possible through *ad hoc* interactions. Moving to an online environment therefore both demands new skills in organisation and formalisation of when and how to interact and familiarisation with the tools of interaction in online environments. These new skills can be taught or can be left for the students to develop for themselves. Both approaches have their merits, although an intermediary position, not of teaching the skills but in teaching critical observation and reflection about the effectiveness of their interactions, may prove a better preparation for learning.
- Monitor and regulate interactions. The degree to which a teacher should be involved in interactions is a debatable point. Learning from experience may provide students with a better learning opportunity to learn how to interact, but if the time taken to learn the essential elements is too long, then this can have a too detrimental result on their collaboration and therefore on the discipline-specific skills that the task has been set up for them to learn. Where intervention is necessary is when their behaviour may be inappropriate, not only with cyberbullying but also more inadvertently excluding those with accessibility issues, for example not switching to text if a member of the collaboration has a hearing impairment.

Online learning

Walton and Hepworth (2011) identified positive changes in cognition associated with experiencing online collaborative learning and put forward a model for a blended teaching and learning intervention (a mix of online and face-to-face approaches) that engages the learner and leads to higher-order thinking. However, as stated, collaborative learning is defined as learning together, and for many "togetherness" is not something that can exist online. The sense of connection and even rapport that occurs between people when sharing a space in the offline world can be replicated in an online environment, but for some students this is difficult and perhaps impossible to acquire, and yet recent studies (Shellenbarger, 2014) indicate the importance of making a social connection when collaborating.

Presence, the experience of "being there" within an online environment (IJsselsteijn, 2005; 8), can be distinguished as either telepresence (experiencing presence at a remote site) or virtual presence (experiencing presence within a computer-generated environment). Within the online interactions discussed within this book, learners are required to interact with others located at remote sites, and this sensation of telepresence has long been part of our experience (Sheridan, 1992, 120).

Building on this sense of telepresence is another element of online experience, that of social presence, defined as the "ability to project oneself socially and emotionally in an online community" (Arbaugh and Hwang, 2006; 10; Caspi and Blau, 2008; 324) and also the ability to perceive others (Becker and Mark (2002; 29). However, the sense of togetherness extends beyond the ability to perceive another to that of experiencing the sensation that everyone is present and co-located within a single shared environment. This is referred to as "copresence" (Zhao, 2003; 445).

A key part of creating this co-presence is to share personal information. The more this is done, then the more likely others are to reciprocate and, consequently, build up trust among the members of the group (Rourke *et al.*, 1999; 55). However, the ability of people to develop these social connections varies widely and is dependent on the skills and abilities at perceiving social nuances within an online environment or even the value and purpose of online interactions (Caspi and Blau, 2008; 339). The development of trust is essential to the effectiveness of online interactions, and yet the predisposition and ability of students to create the social linkages that enable copresence emerge may therefore vary enormously.

Online learners

At the time of writing this book, traditional learners within current undergraduate programmes were born in the mid-1990s. Whereas the so-called Millennials or Generation Y students were entering higher education around the turn of the millennium, the current students are the generation following these, sometimes referred to as Generation Z (Howe and Strauss, 2000; 4). These students have grown up with the existence of the World Wide Web and with social media, and yet observations indicate that this does not mean automatically that these students are adept at using these technologies (Kennedy *et al.*, 2006; Margaryan and Littlejohn, 2008), or their use of technology has a significant impact on the manner in which they learn (Bennett, Maton and Kervin, 2008; 779). As Bennett, Maton and Kervin note, "It may be that there is as much variation within the digital native generation as between the generations" (2008; 779).

The term "digital native" was coined by Prensky in 2001, where he asserts that the students who grew up using digital technologies have actually developed different thinking patterns as a result (Prensky, 2001; 2). According to Prensky, users whose early development was in a period before digital technologies were commonplace may therefore be able to adopt some of the techniques and language but will always use a different language, be socialised differently and even have different modes of thinking. Prensky has since recognised that the dichotomy between natives and immigrants is too simplistic (Prensky, 2009).

A characteristic of previous generations of learners was that, although many developed a sense of copresence online, a minority (usually estimated at between a third and a quarter) do not or cannot. Towell and Towell (1997; 593) reported that 69% of participants in a text-based networked virtual environment experienced a sense of presence, even though the only medium through which they interacted was text. Several users also reported that their sense of presence depended on who they were interacting with and the topic under discussion (1997; 593). Previous studies by the authors indicate that the motivation to learn the subject matter also contributes to a sense of online copresence (Childs and Peachey, 2013; vii).

Bayne (2004), in direct contrast to Towell and Towell's findings, recounted the experience of participants in an online course who failed to experience co-presence throughout their online interactions. Her interviewees' experiences were that the act of seeing body language is so fundamental to communication that without it, the communication fails to seem real. Bayne states that:

> Communicating online is perceived here as being an interpretive act in a sense that intercorporeal communication is not. The loss of the language of the body leaves 'just words'. . . resulting in a communicative act that is 'not real' in the sense that its emotional contexts are purely constructed, a matter of interpretation.
>
> (Bayne, 2004)

Although so-called Generation Y and Z learners grew up in an online environment, as can be seen in the subsequent chapters, this expression of inauthenticity of their online interactions still persists for many students, with a consequent impairment of their ability to collaborate online.

Employability

Employability skills are often defined as the skills that make graduates employable at the end of their study. However, Tymon (2013) argued that the emphasis of employability should be on equipping graduates with lifelong skills to contribute to society at large, rather than just getting "employed" following graduation. This long-term view of employability is in line with the growing recognition and contribution of the "soft" people/management skills and the need to equip graduates with appropriate skills for industry employment in the UK (Spinks, Silburn and Birchall, 2006; RAE, 2007; Lamb *et al.*, 2010) and worldwide (Beanland

and Hadgraft, 2014). In recent years, the architecture, engineering and construction (AEC) industries have been undergoing a transition to digital industry with the emergence of the BIM way of working. A report by BIM2050 group (2014) presented a scenario of this transition through four chronological waves to 2050, which demand different skill sets. The report suggests a diminishing of traditional professional boundaries, preferring graduates with a wider understanding of the processes and multidisciplinary knowledge. Hence, STEM courses are not only expected to offer cross-discipline competence in addition to disciplinary expertise but also should nurture a culture of integration and collaboration. Further, STEM educators should proactively develop an online virtual community of learning (BIM2050 group, 2014) as a support mechanism to facilitate this culture. Skills of online learning/collaboration will not only bring a potential for enhancing employability following graduation but, importantly, will enable the STEM graduates to acquire lifelong skills to work in rapidly changing industries globally.

Remote communication and collaboration

Remote collaboration in building and construction

Remote collaboration between geographically distributed team members is a growing phenomenon in global engineering projects (Kirkman *et al.*, 2002; Gibson and Gibbs, 2006; Nayak and Taylor, 2009). There are two reasons this particular industry makes particular demands on virtual collaboration. One of these is that the increase in complexity in the building process requires collaboration among specialists from a wide range of disciplines, and bringing together architecture, engineering and construction teams into one virtual extended team enables far more effectiveness when team members collaborate and synthesise across knowledge domains (Carrillo and Chinowsky, 2006; Whyte *et al.*, 2008). Since these teams are often multinational in nature, as well as multidisciplinary, this can lead to a wide range of communication issues. Working across national and organisational boundaries also can impede the flow of information. These all work to impede communication, collaborative work and joint problem solving, even when contractual agreements are designed to encourage a collaborative environment (Mitropoulos and Tatum, 2000; Dossick and Neff, 2010).

The second specific issue that construction engineering faces is the combined design and sharing of analysis and problem solving, which is not only complex but requires the synthesising of abstract and conceptual ideas to a degree that even co-located teams struggle to do efficiently and effectively, which makes the challenges facing virtual teams even more difficult to resolve.

Transactional distance

Although the construction and building industries face particular challenges, all of the activities described in this book face the issue of conducting effective global collaboration. Ingram and Hathorn (2004) define collaboration as having three essential elements: participation, interaction and synthesis (the creation of new knowledge). Participation, interaction and synthesis at a distance presents challenges; however, these difficulties do not necessarily differ in *nature* to those encountered when face to face but simply differ in matter of degree. The psychological separation between two people in any dialogue (in the examples given by Moore [1993; 22–24] between tutor and student) are due to differences in language, in culture, in skill at communication, in subject discipline and experiences. The psychological separation that results from these is known as transactional distance (Childs, 2010; 54), and though their mediation via technology may increase this separation, these constraints also exist in face-to-face communication.

Moore (1993; 28–30) adds that by structuring the communication, this transactional distance may be reduced to some extent; this structure can be created through:

- Using multiple media to present information
- Encouraging the motivation of the learner
- Encouraging analysis and critical reflection
- Providing constant feedback and advice
- Facilitating the students' co-creation of knowledge
- Arranging for practice, application, testing and evaluation

The essential element of Moore's position is that promoting effective dialogue achieves reduction in transaction distance, in addition to improving environmental factors such as changing or improving the communication medium. Installing a higher-quality technological platform is only one of many factors. However, this is often the factor focused upon in distanced communication.

Visualisation

As stated, construction engineering creates many challenges, even for co-located teams, through its requirement to bring together many disciplines and resolve many quite abstract problems. Many of these issues can be overcome through visualisations, sketches and models, as these serve as both a way to communicate knowledge and a way to make tacit and knowledge explicit (Whyte *et al.*, 2008). This process is known as reification (Wenger, 1998; 58), which is the process of making abstract concepts concrete, or in, Wenger's phrase, giving them "thingness". They also provide the function of documenting work within the project. Once visualised and modelled, ideas can be shared, understood and mutually worked upon so that differences can be more easily resolved; models and documents become the basis for conversation and from there co-construction of further meaning (Suwa *et al.*, 2000; Ewenstein and Whyte, 2007; Whyte *et al.*, 2008; Neff, Fiore-Silfvast and Dossick, 2010).

Once these visualisations and models have been created they, can be distributed to the virtual teams through knowledge exchange systems. In the construction industry this is known as an important part of the BIM way of working

(Orlikowski, 2000; Taylor, 2007; Whyte *et al.*, 2008). This is not to say, however, that this transmission is entirely transparent; even at this stage it is open to reinterpretations and reframing as the recipients of a drawing or a model make sense of it through their own domain lens, their role on the project and their disciplinary expertise (Dossick and Neff, 2010), let alone their differing interests on the project.

Messy talk

Part of the process of sense making across subject domains, cultures, languages or even simply two different perspectives on a problem involved what Dossick and Neff define as "messy talk". Messy talk is an active, informal and flexible collaboration process and supports brainstorming and mutual discovery (2011; 85). From their observations of co-located teams working with BIM tools, Dossick and Neff (2011) identified the key characteristic of messy talk as being a means for participants in teams to mutually discover issues that their colleagues may not be aware of and need to know. In the "messy talk" phase, the ideas, problems and misunderstandings that occur are unanticipated and emerge haphazardly, which is why formal, structured approaches to communication are not effective at supporting it. The incremental iterative nature of this discovery phase is more likely to be brought about by impromptu, off-topic and "off-the-top-of-the-head" types of comments. For designers particularly it is "The intersection of scope, constraints, constructability, and design intent [that] is often unanticipated" (p. 91).

Dossick *et al.* (2012) define the following four key elements of messy talk as:

- Mutual discovery: The process is one of discovery, both the discovery of solutions and of new problems that emerge from proposed solutions.
- Critical engagement: Individuals actively engage in thinking through the issues and problem solving
- Knowledge exchange: Team members exchange information about a situation that was previously distributed among them – no one member has knowledge or information that leads to the identification of the problem and the solution
- Synthesis: Recognition by the team that the solution is the result of interaction.

Visualisation and messy talk in tandem

Separately, talk and visualisations help to bridge transactional distance. However, Dossick and Neff further argue that it is in combination that these two modes are at their most effective. Visualisations support messy talk by allowing people to draw, write, sketch, talk or otherwise modify shared knowledge together and may lead unexpected discoveries through designers' process of sketching, analysis and synthesis (Suwa *et al.*, 2000; 240). These are particularly valuable in multidisciplinary settings. Dossick and Neff also conclude that while visualisations may support messy talk, teams need to be organised and ready to engage in it in order to generate new ideas or innovate.

The studies cited were all conducted within co-located teams. Working in virtual teams creates a greater transactional distance between participants since all interactions are mediated through technology, and they are also more likely to have cultural and linguistic differences. This separation also makes the factors that promote collaboration (establishing trust, fostering productive informal communication, cultivating knowledge exchange) more problematic. An additional challenge for virtual teams in the AEC industries is that these teams are most commonly organised in temporary multi-organisation teams or project networks and usually supported through digital documentation systems such as BIM.

However, the findings of Dossick and Neff (2011) are that platforms such as BIM may actually constrain the ability to develop understanding between participants because these tools are designed for the efficient sharing of finalised documentation and so limit the opportunities for the "messier" mutual discovery and unanticipated problem solving that has been described. While BIM supports problem definition and explicit knowledge creation, it is less powerful for joint problem solving (Dossick and Neff, 2011). Participation and interaction might be well supported using current BIM and videoconferencing, but achieving synthesis could be more challenging online where the types of informal and flexible side conversations that occur with co-located teams are not so easily conducted.

Conclusions

The disciplines of science, technology, engineering and maths are increasingly collaborative and require collaborative skills in these skills within the workplace. These collaborations are increasingly conducted in distributed teams, and this online collaborative working requires specific competencies to conduct them effectively. As educators in STEM subjects, therefore, we need to prepare our students for the realities of working in these online workplaces, and yet to some extent, we are still identifying what these skill sets are.

Throughout the following chapters, therefore, the following themes emerge and are central to our investigation of how learners collaborate online.

- What skills are required for students to learn and collaborate effectively?
- If collaboration is "learning together", to what extent do students experience togetherness and how can this be facilitated?
- What impact does collaboration have on learning, learning outcome and employability skills?
- How can technology enhance or detract from collaborative learning?
- To what extent do students value online collaborative learning, and what aspects do they find rewarding or daunting?
- What can we learn from the students' experience of online collaboration to enhance our own practice as educators?

Through taking a multifaceted approach to exploring these questions, we aim to contribute to the debate but also present some essential practical aspects of encouraging online collaborative skills in our students.

References

Bayne, S. (2004). *Mere Jelly: The Bodies of Networked Learners, Networked Learning Conference*. www.networkedlearningconference.org.uk/past/nlc2004/proceedings/individual_papers/bayne.htm.

Beanland, D., and Hadgraft, R. (2014). *Engineering Education: Transformation and Innovation. A Monograph Commissioned by UNESCO*. Melbourne: RMIT University Press.

Bennett, S., Maton, K., and Kervin L. (2008). The 'Digital Natives' Debate: A Critical Review of the Evidence. *British Journal of Educational Technology*, 39(5), 775–786.

BIM2050 group. (2014). *Built Environment 2015: A Report on our Digital Future*. London: Construction Industry Council.

Carrillo, P., and Chinowsky, P. (2006). Exploiting Knowledge Management: The Construction and Engineering Perspective. *Journal of Management in Engineering*, 22(1), 2–10.

Childs, M. (2010). Analysis and Description of Education Employing Technological Platforms: Terminology, Features and Models, in Clouder, L., and Bromage, A. (eds.), *Interprofessional E-Learning and Collaborative Work: Practices and Technologies* (pp. 46–60). New York: IGI Global.

Childs, M., and Peachey, A. (2013). *Understanding Learning in Virtual Worlds*. London: Springer.

Childs, M., Peachey, A., Jackson, E., and Hall, P. (2013). Grandfathers, bots and gloop: Learner choices for designs of companion agents. *EDMEDIA Conference*, 24–28th June, 2013, Victoria, British Columbia, Canada.

Cormier, D. (2008). Rhizomatic Education: Community as Curriculum. *Innovate*, 4(5). http://davecormier.com/edblog/2008/06/03/rhizomatic-education-community-as-curriculum/

Dainty, A.R.J., and Edwards, D.J. (2003). The UK Building Education Recruitment Crisis: A Call for Action. *Construction Management and Economics*, 21, 767–775.

Dillenbourg, P. (1999). What Do You Mean by Collaborative Learning?, in Dillenbourg, P. (ed.), *Collaborative-Learning: Cognitive and Computational Approaches* (pp. 1–19). Oxford: Elsevier.

Dossick, C.S., and Neff, G. (2011). Messy Talk and Clean Technology: Communication, Problem Solving, and Collaboration Using Building Information Modeling. *Engineering Project Organizations Journal*, 1(2), 83–93.

Dossick, C.S., Anderson, A., Iorio, J., Neff, G., and Taylor, J. (2012). Messy talk and mutual discovery: Exploring the necessary conditions for synthesis in virtual teams. Working Paper Proceedings, Engineering Project Organizations Conference Rheden, The Netherlands.

Driscoll, M.P. (2005). *Psychology of Learning for Instruction*. London: Pearson.

Egan, J. (2004). *Skills for Sustainable Communities*. London: Office of the Deputy Prime Minister.

Ewenstein, B., and Whyte, J.K. (2007). Visual Representations as 'Artefacts of Knowing'. *Building Research & Information*, 35(1), 81–89.

Forde, C., and MacKenzie, R. (2004). Cementing Skills: Training and Labour Use in UK Construction. *Human Resource Management Journal*, 14(3), 74–88.

Gibson, C.B., and Gibbs, J.L. (2006). Unpacking the Concept of Virtuality: The Effects of Geographic Dispersion, Electronic Dependence, Dynamic Structure, and National Diversity on Team Innovation. *Administrative Science Quarterly*, 51(3), 451–495.

Gokhale, A.A. (1995). Collaborative Learning Enhances Critical Thinking. *Journal of Technology Education*, 7(1).

Howe, N., and Straus, W. (2000). *Millennials Rising: The Next Great Generation*. New York: Vintage/Random House.

IJsselsteijn, W.A. (2005). History of Telepresence, in Schreer, O., Kauff, P., and Sikora, T. (eds.), *3D Communication: Algorithms, Concepts and Real-Time Systems in Human Centred Communication* (pp. 7–22). Chichester: John Wiley & Sons.

Ingram, A.L., and Hathorn, L.G. (2004). Chapter X: Methods for Analyzing Collaboration in Online Communications, in Roberts, T. (ed.), *Online Collaborative Learning: Theory and Practice* (pp. 215–241). Melbourne: Information Science Publishing.

Kafai, Y.B., and Resnick, M. (2011). Introduction to Constructionism, in Kafai, Y.B., and Resnick, M. (eds.), *Constructionism in Practice: Designing, Thinking and Learning in a Digital World* (pp. 1–8). New York: Routledge.

Kennedy, G., Krause, K.-L., Judd, T., Churchward, A., and Gray, K. (2006). First Year Students' Experience with Technology: Are They Really Digital Natives? www.bmu. unimelb.edu.au/research/munatives/natives_report2006.pdf.

Kirkman, B.L., Rosen, B., Gibson, C.B., Tesluk, P.E., and McPherson, S.O. (2002). Five Challenges to Virtual Team Success: Lessons from Sabre, Inc., *The Academy of Management Executive (1993–2005)*, 16(3), 67–79.

Krathwohl, D.R. (Autumn, 2002). A Revision of Bloom's Taxonomy: An Overview. *Theory into Practice*, 41(4), 212–264.

Lamb, F., Arlett, C., Dales, R., Ditchfield, B., Parkin, B., and Wakeham, W. (2010). *Engineering Graduates for Industry*. London: The Royal Academy of Engineering.

Margaryan, A., and Littlejohn, A. (2008). Are Digital Natives a Myth or Reality? Students' Use of Technologies for Learning. www.academy.gcal.ac.uk/anoush/documents/ DigitalNativesMythOrReality-MargaryanAndLittlejohn-draft-111208.pdf.

Mitropoulos, P., and Tatum, C.B. (2000). Forces Driving Adoption of New Information Technologies. *Journal of Construction Engineering and Management*, 126(5), 340–348.

Moore, M.G. (1993). Theory of Transactional Distance, in Keegan, D. (ed.), *Theoretical Principles of Distance Education* (pp. 22–38). London: Routledge.

Nayak, N.V., and Taylor, J.E. (2009). Offshore Outsourcing in Global Design Networks. *Journal of Management in Engineering*, 25(4), 177–184.

Neff, G., Fiore-Silfvast, B., and Dossick, C.S. (2010) A Case Study of the Failure of Digital Media to Cross Knowledge Boundaries in Virtual Construction. *Information, Communication & Society*, 13(4): 556–573.

Orlikowski, W.J. (2000). Using Technology and Constituting Structures: A Practice Lens for Studying Technology in Organizations. *Organization Science*, 11(4), 404–428.

Prensky, M. (2001). Digital Natives, Digital Immigrants, On the Horizon, 9(5), October 2001, NCB University Press. www.marcprensky.com/writing/Prensky%20-%20Digital %20Natives,%20Digital%20Immigrants%20-%20Part1.pdf. [Accessed 10 February 2007].

Prensky, M. (2009). H. Sapiens Digital: Digital Immigrants and Digital Natives to Digital Wisdom. *Innovate*, 5(3), www.innovateonline.info/index.php?view=article&id=705. [Accessed 4 February 2009].

RAE. (2007). *Educating Engineers for the 21st Century*. London: The Royal Academy of Engineering.

Sheridan, T. (1992). Musings on Telepresence and Virtual Presence. *Presence: Teleoperators and Virtual Environments*, 1(1), 120–126.

Spinks, N., Silburn, N., and Birchall, D. (2006). *Educating Engineers for the 21st Century: The Industry View. A Study Carried Out by Henley Management College for the Royal Academy of Engineering*. Henley-on-Thames: Henley Management College.

Suwa, M.J., Gero, et al. (2000). Unexpected Discoveries and S-Invention of Design Requirements: Important Vehicles for a Design Process. *Design Studies*, 21(6), 539–567.

Taylor, J.E., and Levitt, R. (2007). Innovation Alignment and Project Network Dynamics: An Integrative Model for Change. *Project Management Journal*, 38(3), 22–35.

Towell, J., and Towell, E. (1997). Presence in Text-Based Networked Virtual Environments or "MUDS". *Presence*, 6(5), 590–595.

Walton, G., and Hepworth, M. (2011). A Longitudinal Study of Changes in Learners' Cognitive States During and Following an Information Literacy Teaching Intervention. *Journal of Documentation*, 67(3), 449–479.

Wenger, E. (1998). *Communities of Practice: Learning, Meaning and Identity*. Cambridge: Cambridge University Press.

Whyte, J., Ewenstein, B., Hales, M., and Tidd, J. (2008). Visualizing Knowledge in Project-Based Work. *Long Range Planning*, 41, 74–92.

Wilson-Medhurst, S., Dunn, I., White, P., Farmer, R., and Lawson, D. (2008). Developing activity led learning in the faculty of engineering and computing at Coventry University through a continuous improvement change process. *Proceedings of Research Symposium on Problem Based Learning in Engineering and Science Education*, Aalborg University, June 30–July 1.

Author biographies

Mark Childs is a Senior Lecturer in Technology Enhanced Learning at Oxford Brookes University, UK, supporting staff in developing their digital capabilities. Mark has been working in higher education since 1995 as a researcher in the use of technology for learning and teaching. Mark's main research area has been the study of learners' experiences of synchronous technologies in higher education and the role of presence and embodiment in supporting learning in these. His PhD in education was completed in 2010.

Robby Soetanto is Senior Lecturer in Construction Management at Loughborough University, UK. Prior to this, he held post-doctoral research posts at Wolverhampton and Loughborough Universities and lectured at Coventry University. His research has been funded by government bodies (EPSRC, EU, HEA, British Council) and private companies (Lloyds TSB and Hewlett Packard). He led Learning to Create a Better Built Environment and BIM-Hub projects, which focus on developing virtual collaboration skills of built environment students and provide the basis of the first part of this book.

Zulfikar A. Adamu is a Lecturer in Architectural Technology at Loughborough University, UK. His research and expertise centres around building information modelling (BIM). He is a BRE certified BIM Accredited Professional (BIM AP).

Stephen Austin is Head of School – Energy, Construction and Environment at Coventry University, UK. He is interested in the development of student professional attributes including communication, leadership and teamwork through integrated project work.

Tomasz Arciszewski is Professor Emeritus and Inventive Engineering scholar at George Mason University, USA. Published books include *Successful Education. How to Educate Creative Engineers* and *Inventive Engineering. Knowledge and Skills for Creative Engineers*.

Arosha K. Bandara is the lead educator for the Open University's Introduction to Cyber Security MOOC. His research focuses on engineering adaptive security and privacy mechanisms into ubiquitous computing systems.

Helen Caldwell is a Senior Lecturer in teacher education at the University of Northampton, UK, where she is curriculum lead for Primary Computing. Her research focuses on the role of professional learning communities in technology education.

Carrie Sturts Dossick is a Professor of Construction Management at the University of Washington, USA, in the College of Built Environments and the executive director of the Center for Education and Research in Construction (CERC).

Jacqueline Glass is Associate Dean for Enterprise in the School of Civil and Building Engineering at Loughborough University, UK. Her research and expertise are in sustainable construction and building design, and she teaches construction and design management students. Her specialism is in responsible and ethical sourcing of construction products.

Jack Steven Goulding is a Chartered Construction Manager, Professor of Construction Project Management and director of the Centre for Sustainable Development (University of Central Lancashire, UK). He is also the co-editor of *Construction Innovation.*

Chinwe Isiadinso is a Teaching Fellow in the School of Civil and Building Engineering at Loughborough University, UK. Her responsibility is currently as a business and strategic manager within Ace Development Consulting (ADC) Ltd. Her research interests relate to sustainable design and tools/mechanisms which enhance the performance of green buildings.

Peng Li received his PhD in electrical engineering from University of Connecticut, USA. He is an Associate Professor in College of Engineering and Technology at East Carolina University, USA.

Ken-Yu Lin is an Associate Professor of Construction Management and also the director for the Construction Management Occupational Safety and Health (CMOSH) program at the University of Washington, USA.

Helen Mackenzie is an independent research consultant. She is particularly interested in undertaking qualitative research to examine different students' personal experiences of "transitional space" within higher education.

Giovanni C. Migliaccio is an Associate Professor of Construction Management at the University of Washington, USA, and a Fellow with the Runstad Center for Real Estate Studies and holds a P.D. Koon Endowed Professorship in Construction Management.

Paul S.H. Poh teaches Construction Project Management at Ryerson University in Toronto, Canada. He has a wide range of industry experience – in a consultant firm and in construction as well as for a public authority.

Farzad Pour Rahimian is a Senior Lecturer in Building Information Management at the Department of Architecture, Faculty of Engineering, University of Strathclyde, UK. He is also the co-editor of the MIT–based *International Journal of Architectural Research.*

Mike Richards is a Senior Lecturer at the Open University, where he teaches an introduction to ubiquitous computing, as well as co-author of a highly successful FutureLearn course on cybersecurity. When not teaching, he likes to play with volcanoes.

Neil Smith is a Senior Lecturer in the Computing and Communication department at the Open University. He leads the outreach activities for the department. He has worked with Code Club and Computing At School to develop materials for use by students and teachers for computer science education. His research is in computer science education and artificial intelligence.

Harry Tolley is a research consultant. He was an Honorary Professor (University of Nottingham, UK) and External Evaluator to the Engineering Subject Centre and engCETL (Loughborough University, UK).

Part I

Lead case study

International collaboration for multidisciplinary built environment education

1 International collaboration for multidisciplinary built environment education

Robby Soetanto, Mark Childs and Stephen Austin

Introduction

Part One of this book details the findings from two interlinked projects. The first of these ran from 2010 to 2012 and was funded by the Hewlett Packard Catalyst Program. The title of this project was Learning to Create a Better Built Environment, and it consisted of a collaboration between Coventry University (CU) in the United Kingdom and Ryerson University (RU) in Toronto, Canada.

In July 2013, these collaborators, together with the addition of Loughborough University, also in the UK, were awarded funding by the UK's Higher Education Academy to continue this work with a follow-on project. The project (BIM-Hub) continued until March 2015.

The following chapters in Part One detail the findings of these projects; to provide a context for these findings, this chapter describes how the projects ran and the external factors which led to these projects being constructed.

Context for the learning to create a better built environment project

The first of the two projects was influenced by a number of parties inside and outside participating universities. Being final-year students, participating students had experienced learning within the context of a working group. They also learnt people skills in the management of group work and team dynamics. However, this was a major departure from the normal practice for group work within the universities, and so, although the direct line management within the universities (i.e. heads of department) were very supportive of learning innovation, as it aligned well with the current campaign for applying innovative learning in the faculties, they were a little nervous of the impact on the final year students. The concern was that if this 'experiment' did not go according to plan, it might create resentment from the students. This was particularly true in the UK given the increasing emphasis of the National Student Satisfaction survey (NSS), which informs the national subject ranking (e.g. league tables) and which therefore has an impact on future student recruitment. To mitigate this potential issue, the tutors attempted to balance the workload and experience between participating and non-participating students.

Outside the university, this innovative learning was designed to meet the expectation of employers (construction companies, consultants, architects etc.) for the internationalisation of education in the built environment (BE). The courses/programmes in the two institutions were accredited by professional institutions in the UK and Canada and therefore had to meet their accreditation requirements. In the existing educational system, activities in the courses should be justifiable to external examiners from other universities. Members of advisory boards of the courses could have an impact on the delivery of the project. The project existed due to sponsorship from Hewlett Packard Catalyst Initiative and therefore determined the way the outcomes of learning were collected and examined. With the expectation that the project continues in the future, there was a need to identify potential future sponsors for the activities. The innovativeness of the project would also provide a good input for the course marketing team to talk to prospective students about the innovative approach of learning in the institutions.

As a proactive response to the increasingly competitive educational environment and a shift in modern educational agenda, a new pedagogical approach, based on problem-based learning principles, was implemented within the department at CU. The aim was to promote student engagement, retention and employability. In addition, part of the university's mission statement was to "aspire to be a dynamic, global, enterprising university". By aspiring to be a global university the university aims to "ensure that every course contains a substantive international element that requires students: to conduct joint projects with peers in another university overseas mediated via digital technology". International mobility was seen as critical to the success of construction business, both today and in the future. An increasingly interconnected and globalised construction sector means that individuals often have to work collaboratively with people in different parts of the world who they have not even met before, let alone worked with. Construction work, by the very process it undertakes to produce the final product, tends to be formed from multidisciplinary teams. These multidisciplinary teams can bring innovative products and processes aimed at benefitting the construction process. However, effective teamwork skills in terms of management, leadership and communication are needed if construction projects are going to meet their prescribed output. This was confirmed by a UK government-commissioned review on the skills for creating sustainable communities highlighting the importance of communication and professional skills for built environment graduates (civil and building engineering), which require further development (Egan, 2004). Therefore, the activities described in Part One of this book were intended to implement the innovative pedagogical approach whilst addressing the increasing need for student internationalisation within the construction industry.

The study design

The study was undertaken within the context of international institutions, which educated students for future careers in the BE sectors. The main focus was the final-year undergraduate students, although some aspects of the study were conducted with the MSc postgraduate students. From the UK institution, in the first round, 35 out of 249 undergraduate students participated voluntarily in the study. This was closely matched by the number of students from Canada: 37 students from the entire cohort. Participating UK students studied civil and/ or structural engineering, whereas Canadian students studied architecture. One team typically comprised an equal number of students from both institutions, four from UK and four from Canada. Two hundred and fourteen non-participating UK students who worked on the same building project but with their co-located team provide a comparative 'control' group for analysis at later stages of the project. Studying in international institutions, they comprise a mixture of ethnic and educational background with English being their first or second language. Most of them were in their early 20s, and only few really had real industry experience.

The investigation at postgraduate level included students with full- and part-time modes who participated in one module. They were assigned a piece of group coursework and submitted a critical assessment of how their group work was mediated by communication technology, including comments that compared the collocated and distanced collaborating groups.

The students formed multidisciplinary groups to design a new academic departmental building. A project scenario was developed based on a real academic building, which would be built in the future to replace the existing building. The project lasted for one academic year. This design project represents a 'problem' to be solved by students in groups, and this mimics an authentic industry practice in the real working world. It was considered an integration of theory and knowledge that they obtained earlier for practical application within a real-world context. Apart from the application of the knowledge and skills within the specific subject domains, the work in the project was intended to develop the students' 'soft' skills, including people management (face to face or virtual), communication, teamwork and time and self-management, which are considered important for future employability and increasingly being emphasis by the employers in recruitment. The project was designed for a more engaging mode for learning.

A project briefing was devised that allowed the teams the opportunity to successfully work together within the function of a design-and-build context. Academics from both institutions were involved in the development of the brief to ensure that it met the learning outcomes of their respective degrees, namely to "demonstrate integration within a group of various construction professions, previously learnt knowledge and skills for a major scenario-based project, within a design and construction environment" and to "demonstrate a critical approach to skills through teamwork and continuous personal improvement". The project briefing was communicated early in the academic year before the formation of the teams. The comprehensive project brief included (i) description of purposes of building (i.e. scenario), requirements of facilities (e.g. rooms, area, environmental aspects), site location and constraints (relationships with the existing building and facilities in the surrounding area) and schedule of accommodation, (ii) requirements on group formation and work processes – team leadership, management and documentation (meetings, roles of individual student) and (iii) assessment of tasks with detailed requirements for two project phases, and peer assessment using the WebPA system (see Wilkinson and Lamb, 2010 for description on WebPA). In addition to these, design guidance of building standards, structural design codes, photographs of the site and poster and presentations were also provided.

The brief was fully reviewed by academic teams from both universities to ensure that the tasks that the students were being asked to complete would be at the appropriate academic level and also offer a range of multidisciplinary tasks to the virtual teams to undertake. The project was split into two phases to coincide with academic year structures: phase 1 – September to December and phase 2 – January to April. The multidisciplinary tasks for phase 1 (design) were to architecturally and structurally design the proposed building including parking, drainage and accessibility, provide an initial cost estimate and outline specification. The tasks for phase 2 (tender) were to finalise the architectural and structural design including parking, drainage and accessibility, provide reports on quality and health and safety and create a programme for the construction of the project.

Local groups of four students were formed in the participating universities. The teams reviewed the tasks in the project brief and identified previous technical skills to meet the tasks. The teams then linked with another team from the other institution to form one virtual team of eight students (four from the UK and four from Canada). As the teams had not worked together before, they needed a way of getting to know each other professionally. This was undertaken through a poster presentation. They were asked to identify people management skills (e.g. leadership, teamworking, communication) that they could demonstrate. Each group was asked to produce one A2-sized poster which should contain technical and management skills of the team, with a view of attracting offers from counterpart teams. The electronic exchange of posters was coordinated by academics from both institutions, who were available to oversee this process and mediate where there were any problems with team choices. The aim was to form the strongest team. Evidence from previous experience was also included in the poster. The teams reviewed the different team posters with a view to negotiating and agreeing with a counterpart team for the formation of a company. The whole exercise was aimed at developing comprehensive understanding of the project brief and reviewing potential strengths and weaknesses of the team members. This exercise developed their skills for identifying expertise to complement the existing team members. Having formed the virtual teams, the students started working on the project. To help the initial process of communication, the first task focused on team formation and planning of the main technical outputs. The 'companies' were asked to plan the forthcoming technical tasks and write reports showing the interdependency within the tasks and how they intended to manage the team. This then led into the specific technical tasks that the students undertook.

An important emphasis of the project was communication that the students needed to undertake as distributed teams. The teams conducted weekly meetings and appointed a company leader and secretary to be rotated every four or five weeks, thus enabling each member of the team to carry out each role. The team leaders chaired the weekly project meeting, monitored and co-ordinated the work of the group, ensured that submission dates were met and generally oversaw the day-to-day running of the project team. The company secretary took the meeting minutes, noting any important points discussed, and deputised for the group leader in the event of their absence. Copies of the meeting minutes formed part of the assessment process. Companies were encouraged to also meet outside the scheduled meeting time. In addition, they were encouraged to use a range of ICT to successfully communicate, including Skype, Messenger and Dropbox for document storage. Academics from both institutions held tutorials with the virtual teams through these forms of ICT to help facilitate any questions regarding the technical tasks and monitor team performance. The marking scheme combined individual and group marks for each task. The individual marks were derived from the assessment of the task that the individual was responsible for. The group mark was derived from the team formation process and presentations. The group mark was peer assessed using the WebPA system. The system provides a control mechanism to discourage students being 'passengers' in the team. Further pedagogical benefits from peer assessment to the skills formation in group work is explained in Wilkinson and Lamb (2010).

The 'companies' needed to submit one document, in response to the technical tasks, at the same time. In addition, the 'companies' needed to liaise with each other and produce one presentation. The presentation was delivered to the academics from the respective university they derived from. The presentation needed to outline the work undertaken in the first phase of the project by the team. When presenting, the teams concentrated on outlining in depth the specific parts they had undertaken but also needed to demonstrate an acceptable level of understanding and comprehension of what the other half of the virtual team had produced. This was particularly relevant to the architectural and structural design of the proposed building. In addition, the 'companies' participated in the department exhibition, which was attended by invited industry professionals and colleagues.

Technologies used in the study

Participating students used computer-aided design (CAD) software for technical drawing and communication technologies. The most popular CAD software was Autodesk Revit Architecture Suite and Autodesk Revit Civil, which allowed the best-performing groups to implement a basic building information modelling (BIM) process. The majority of other groups used Autodesk 2-D/3-D CAD design software, whilst a few groups performed the technical drawing through SketchUp. Email, by far, was the most popular means of asynchronous communication. However, the majority of groups saw the need for visual communication as part of the 'soft' skills they were developing in relation to the module learning outcomes. For this process they mostly used Skype. The Skype meetings tended to take place on a weekly basis with the communication method outside of these meetings being email. Students were given access to University laptops provided by the project funder, Hewlett Packard, to undertake these Skype meetings. They also used Dropbox for depositing files with large size. Other similar 'cloud' storage facilities were also used, but Dropbox was the most dominant. These are operated within the university Internet infrastructure. In the early stage of developing their posters, the students again used a range of software. The majority were produced in Microsoft Word, Publisher or PowerPoint. However, a significant number were produced in Adobe Photoshop, again operated within the university's computer infrastructure. Only one group produced the traditional manual 'cut-and-paste' process. As the posters were required in A2, the groups were required to print out the posters using a colour plotter, provided within the departments.

The assessment of student group work/participation was facilitated by the WebPA system within the university. WebPA allows assessment of individuals by their peers (other members of the group) based on pre-determined criteria. The outcome was a percentage of the group mark that should be allocated to an individual mark.

The implementation of the study

Earlier in the process, aligning modules (or courses) in two institutions presented a significant challenge for the tutors. Aligning processes requires flexibility and compromise to adjustments in order to make this work. It took a few months lead time for negotiation and adjustments from both sides. The requirements (in terms of e.g. commitment, marking schemes etc.) had to be communicated in advance of the implementation of the activity. As previously mentioned, in the UK management were supportive but nervous about the impact of the project due to concerns raised regarding the possible student experience and whether that would affect their student satisfaction level. To try and reduce this possible impact, the UK students who undertook the project were volunteers. They were assured that participating in the activity would not disadvantage them by factoring in possible problems arising from distance collaboration. The academics' proactivity identified and responded to problems raised by the students. To gauge their experience, all UK students undertaking the module were asked to provide feedback based on a module evaluation questionnaire. The results were actually better than at the same time the previous year. Key indicators showed that overall satisfaction for the module had actually increased by 16 to 18%. In addition, a separate questionnaire was undertaken at the end of the project for students working in international teams, focusing on the group's vision, collaborative attitude and learning style.

As the UK students were volunteers, it was also decided that their timetabled lecture would be at an earlier time to the other students undertaking the module. This would allow for a greater level of lecturer/student interaction and also deal with any specific issues affecting the international groups. The separate lecture also allowed the groups the opportunity to use the later timetabled lecture to visually communicate with the overseas team, taking into consideration the 5-hour time difference. The 5-hour time difference was a convention that the students had to work within. Students had to deal with the issue that the preferred form of communication usually resulted in a time-delayed lag in responses. Students therefore needed to appreciate when it would be reasonable to receive a response when asking each other questions. This was particularly pertinent when the clocks were changed to compensate for winter. Canada alters their clocks earlier than the UK. This resulted in a number of groups missing pre-arranged Skype meetings, as the time difference between the two time zones had altered.

Part of the emphasis of the project was the communication aspect that the students needed to undertake as virtual teams. To help the initial process of communication, the first task focused on team formation, planning of the main technical outputs and appointing a team leader and secretary. Tasks were allocated within the groups based on previous technical knowledge. A series of set tasks and optional tasks needed to be completed. Groups chose the optional tasks based on the technical knowledge skill sets within the team. The 'companies' were asked to plan the forthcoming technical tasks and write reports showing the interdependency within the tasks and how they intended to manage the team. This then led into the specific technical tasks that the students undertook. The marking scheme for the project combined group and individual marks. The group mark was derived from the team formation tasks, presentations undertaken and overall group report structure and cohesion. The group grade was then peer assessed. The individual mark was derived from the grade awarded to the technical task or tasks that the individual was responsible for in each phase of the project. Some teams allocated members to more than one technical task. As a consequence a mechanism needed to be introduced to cover this aspect. At the end of each phase of the project the group was required to complete a task allocation sheet stating the individual student's contribution towards each technical task as a percentage. Students who allocated 100% of their time to one specific task were awarded the grade for that particular technical task as their individual grade. The individual grade for students who allocated time to more than one specific task was calculated by using a percentage combination from each of the grades awarded for the different technical tasks.

The tasks that the teams had to undertake included a range of technical focus. To ensure that there was adequate technical support for the students, there was a team of 12 lecturers, across both institutions, to provide guidance to the student teams. Each lecturer had lead responsibility for one of the tasks, aligned to their professional expertise. Student team members were given access to all lecturers to seek guidance on the technical outputs, either verbally or more usually through email for overseas team members. In addition, within the overall team of 12, one lecturer in each institution was responsible for liaising with the other overseas academic to answer queries from within the academic team and co-ordinate the information provided to the student teams. This lead academic was also responsible for developing the 'soft' skills within the project. Through guidance the teams were introduced into a series of team skills including communication, facing risk, leadership, project success, software tools, styles of thinking and production of minutes.

A key role within the project took place at the start of the second phase. Team members were asked to reflect as a group and then as individuals on their performance within the first phase of the project. At the end of the first phase and as part of the WebPA peer assessment, the students were asked to rate themselves and team members against a range of criteria that had been devised by the cohort of students undertaking the project. Once the WebPA had been completed the students were able to access a report which indicated their criteria strength and area for development. Having obtained the report the first meeting of the second phase of the project required an initial group discussion using the results from peer assessment. Members of the group were asked to discuss team members' performance in terms of strengths and areas that could be improved, with an overall aim of facilitating individual reflection that was to be implemented for the second phase of the project. The individuals then produced reflective commentaries on their performance for phase 1 of the project and highlighted any steps, both positive and negative, that they felt they needed to undertake within the second phase of the project.

Context for the BIM-Hub project

After the end of the Learning to Create a Better Built Environment project, one of the project team moved to Loughborough University (LU). This provided the opportunity to revisit and expand upon the earlier project. Another development after the original project was the increasing interest in integrating building information modelling (BIM) to taught modules (Adamu, 2014; 4). A particular issue was the skills requirements to participate in this emergent mode of working in the BE sector (BIM 2050, 2014; 2). In this context, BIM is seen as a shared transactional digital environment in which building models and documents can be shared and annotated online.

> BIM models are among the primary outcomes of the BIM process. They are typically graphic 3D models produced by building design and analysis software and make it possible to visually or analytically explore the essential features/functions of a digital representation of a building. With the addition of non-3D aspects such as time (4D), cost (5D) and post-occupancy management (6D), BIM is applicable to the entire lifecycle of a building.
>
> (Adamu, 2014; 7)

The advanced BIM way of working is intended to enable people collaborating within BIM to simultaneously view, manipulate and develop 3D models of buildings. Not only does this enable a real-time co-creation of building models, this 3D model enables joint analysis and discussion of the building and acts as a platform for overlain content such as links to a construction programme, costs of items and additional metadata, added at both the pre-occupancy and post-occupancy stages. The introduction of a collaborative platform across the BE sector provided a new currency for the aims of this new project, in that this meant that the skills required to effectively collaborate within such an environment would also need to be introduced and developed.

The BIM-Hub project aims to explore an innovative approach to international collaborative design project using a real-time online platform in order to enhance the employability skills. The project was funded by the UK's Higher Education Academy and began in September 2013. It ended in March 2015, with the final report submitted to the funding body in August 2015. The key output of the project was a guidance document, which describes experience, mini case studies, lessons learnt and effective practices (included in the project website at http://bim-hub.lboro.ac.uk/).

Including an additional university in the collaboration meant that the collaborative teams consequently comprised three pairs of two students. The LU students were studying construction engineering management, and in a further additional variation, they were not studying the module linked with the project for both semesters, unlike the students at RU and CU. This meant one pair of LU students joining the collaborative team for the first semester, which were then replaced by a different pair of students for the second semester.

Similar to the previous project, the students were asked to submit a reflective assignment at the end of their involvement with the activity. This provided an additional and very rich source of data, as well as proving to be a rewarding learning experience for the students.

The activity design was kept the same in principle and consisted of creating a building design for the first semester then submitting a report on this design for the second. As with the first project, students were advised to identify a platform for communication and for sharing documentation and were given an open choice on which platform to use. For their synchronous communication they were instructed to use GoToMeeting. Students were also asked to submit a GoToMeeting recording of one of their meetings. This was not to be assessed but was only to be used as observational evidence of their interactions and to provide valuable input for case studies and guidance document as project deliverables. Given the shorter duration that the students collaborated, the team formation facilitated by poster presentation was not undertaken. Instead, groups were allocated by the tutors. The BIM-Hub saw the development of a multi-institutional peer-assessment system based on WebPA and Moodle to support the delivery of the modules across three institutions. The students in a group submitted a single report by uploading the files to Moodle.

Some earlier findings from learning to create a better built environment and BIM-Hub projects

Understanding influencing factors and their interrelationships in a collaborative design project is considered the first step to improve the effectiveness of learning. A conceptual model of virtual collaborative learning was developed by Soetanto *et al.* (2014) through interviews with 23 students who were members of seven building design teams and is presented in Chapter 2, Figure 2.1. The model was underpinned by theory of transactional distance and used to group the factors under sequential input, process and output categories. The factors may influence the process and represent inputs and outputs of virtual collaboration in a building design project. It was found that that trust and professional ethos were the most influential factors in successful team collaboration. The root cause of the breakdown in trust was usually the failure to produce work on time. The success and failure in development of trust can been illustrated in two cycles, one with increasing collaboration and commitment and the other with decreasing collaboration and commitment, as presented in Figures 2.2 and 2.3. Chapter 2 presents detailed analysis of student experience of the Learning to Create a Better Built Environment project. The research evaluated the impact on student performance and benefits. The students believe that the virtual collaborative design project has enriched their experience of working with international partners and will benefit their CV (i.e. employability). This complements the analysis of the assessment results as reported in Soetanto *et al.* (2014) that the project has no impact on individual and group marks, but it has developed a proactive attitude among the participating students. This finding aligns with one of the key attributes for employability, 'proactive personality', which was highlighted by Tymon (2013).

Building from Soetanto *et al.* (2012, 2014), Soetanto *et al.* (2015) investigated key factors related to trust and collaboration and learning styles in virtual design teams based on 192 completed questionnaires of participating students, obtained in pre- and post-implementation stages in two academic years. The findings suggest the need to address perceived risk of distance collaboration in virtual teams for enabling successful collaborative design projects. Lower perceived risk of collaboration will facilitate teamwork and development of trust, but this can only materialise if there is a spirit of collegiality, developed from frequent in-person meetings in a group learning setting (such as a design studio). The research also found differences in preferred communication modes between group members from different disciplines, with the architects tending to prefer visual and kinaesthetic modes and the civil/structural engineers preferring aural and read/write modes. The findings of perceived risk and communication mode suggest the potential influence of disciplinary training to the successful outcome of virtual collaboration in design team. Familiarity with team-based collaboration and studio-based learning, which tends to have higher frequency of in-person meetings, could facilitate the development of trust between colleagues and reduce perceived risk. In this context, it is suggested that tutors should consider mechanisms and activities to reduce the perceived risk of collaboration by, for example, providing guidance, taking on a mediating role (if there is a need for intervention) and encouraging social gatherings or the use of social media (if in-person meetings are impossible). Based on the analysis of group meeting recording, focus group interviews and personal reflections, Chapter 3 describes the student experience, which was then categorised under several dependent factors for developing effective online collaborative working (see Figure 3.1). This finding was used as a guide for implementing collaborative design project (presented in the project website http:// bim-hub.lboro.ac.uk/), which could help alleviate perceived risk of collaboration, facilitate successful collaboration and provide lessons learnt for educators who wish to implement virtual collaboration in their curricula.

In summary, the research has improved an understanding of virtual collaborative design activities through the development of a conceptual model to help to understand the relationship of various factors, the evaluation of impact of the activities on student performance and the development of guidance for virtual collaborative learning. However, there is a need to explore the impact on student employability skills within the context of learning outcomes of professional institutions. Perceived self-efficacy of participating students was also considered an important part of the evaluation. The findings are presented in Chapter 4 of this book.

Summary

This chapter has detailed the context, study design, technologies used and implementation of two projects on international collaborative building design activities, before the earlier research findings of the two projects and their connection to the chapters presented in this book are explained. Further detailed findings are presented in Chapters 2, 3 and 4. Chapter 5 discusses an evaluation from the perspective of the tutors involved.

References

Adamu, Z. (2014). *The Integration of BIM into Taught Modules, Internal Report to the School of Civil and Building Engineering (SCBE)*. Loughborough: Loughborough University.

BIM 2050. (2014, September). *Built Environment 2015: A Report on Our Digital Future*. London: Construction Industry Council.

Childs, M., Austin, S., Soetanto, R., Glass, J., Adamu, Z., Isiadinso, C., Poh, P., Knyazev, D., Tolley, H., and MacKenzie, H. (2014). Virtual collaboration in the built environment. *European Distance and E-learning Network (EDEN) Annual Conference*, 10–13 June, Zagreb, Croatia.

Egan, J. (2004). *Skills for Sustainable Communities: Office of the Deputy Prime Minister*. London: *Her Majesty's Stationery Office (HMSO)*.

Poh, P., Soetanto, R., Austin, S., and Adamu, Z. (2014). International multidisciplinary learning: An account of a collaborative effort among three higher education institutions. *The 8th International Conference on e-Learning*, 15–18 July, Lisbon, Portugal.

Soetanto, R., Childs, M., Poh, P., Austin, S., and Hao, J. (2012). Global Multidisciplinary Learning in Construction Education: Lessons from Virtual Collaboration of Building Design Teams. *Civil Engineering Dimension*, 14(3), 173–181. ISSN 1410–9530 print / ISSN 1979–570X online.

Soetanto, R., Childs, M., Poh, P., Austin, S., and Hao, J. (2014). Virtual Collaborative Learning for Building Design. *Proceedings of the Institution of Civil Engineers – Management, Procurement and Law*, 167, MP1, 25–34. http://dx.doi.org/10.1680/mpal.13.00002.

Soetanto, R., Childs, M., Poh, P., Austin, S., Glass, J., Adamu, Z.A., Isiadinso, C., Tolley, H., and MacKenzie, H. (2015). Key Success Factors and Guidance for International Collaborative Design Project. *International Journal of Architectural Research (ArchNet-IJAR)*, 9(3), 6–25. http://archnet-ijar.net/index.php/IJAR.

Wilkinson, N., and Lamb, F. (2010). *WebPA Online Peer Assessment: Resource Pack*. Loughborough: Loughborough University.

2 Virtual teamworking with undergraduates and postgraduates in the Learning to Create a Better Built Environment project

Mark Childs

Introduction

The Learning to Create a Better Built Environment project (2010 to 2012) involved undergraduates from both Coventry and Ryerson Universities collaborating in teams to create building designs (see the previous chapter). In parallel to this, the opportunity was taken to use many of the techniques involved in the project to support a postgraduate course taking place at Coventry University.

This chapter details the learner experience of the Learning to Create a Better Built Environment project and also reports on the experience of the postgraduates. These experiences differed considerably, and this comparison is examined to identify some of the factors that influence students' participation in online activities.

The data analysis of the undergraduate experience

Focus groups were conducted with students attending Coventry University, one with each group that had taken part in the activity. The interviews were transcribed and separate quotes from the students coded according to three types of responses.

The first set of these responses were those that made reference to barriers, differences or distances between the groups working at the near end (Coventry) and those at the far end (Ryerson). This is based on the notion of transactional distance (Childs, 2010; 54), which is that the psychological separation between two people in any dialogue (in the examples given by Moore [1993; 22–24] between tutor and student) can be described as a series of constraints of which the technology and geographical distance are only additional elements when that communication is mediated via technology. The position of transactional distance theory is that many of these constraints exist in face-to-face communication, such as the personalities and philosophies of the participants, their skill at communication and the content of the dialogue; the technological constraints noticed in distanced communication are merely those that tend to be focused upon by observers and are not necessarily the dominant ones.

The phrases in the second set were coded under the heading of "alignment". Alignment is a concept introduced here as describing the sets of behaviours and activities that act to bridge this separation: either the students' observations of how the two groups are aligned or the process by which they brought the two groups into greater alignment, so, for example, technology may interpose additional distance because students are unfamiliar with it, but this transactional distance can be reduced through practice. The third set of phrases were coded under the heading "Impact"; this is the students' view of the value of the activity to them.

These three elements in combination form the basis of the input/output model that formed the structure of our analysis of the process of online collaboration (see Figure 2.1, Soetanto *et al.*, 2012; 11).

Further to these categories is an additional category of "professionalism". As the data were analysed this aspect emerged as the factor that had the largest influence on the interactions of the students, particularly as it led to the development or erosion of trust. Students' degree of trust led to two divergent experiences, in two different positive feedback loops, in a manner that was not displayed by other distancing and alignment factors and in fact limited or exaggerated the effect of these other factors. Analysing professionalism (and hence trust) separately therefore enabled this specific trait to be compared across the students' engagement and led to the development of a model that summarised these experiences.

Finally, the groups were assigned the categories "successful" and "not successful" collaborations, with the intention of observing differences in behaviour and practice between the two types of groups. In reality the collaborations did not sit at the end of these two polarities but on a continuum. The selection was based on two criteria, that the students themselves identified particular issues with their working relationship with the other team and which predominated: the number of quotes referring to distances and differences or those that referred to alignments. In all, groups A1, A4, A5 and A7 were characterised as "successful" collaborations and groups A2 and A3 as "not successful". Group A6, lying in the middle of these, was characterised as partially successful.

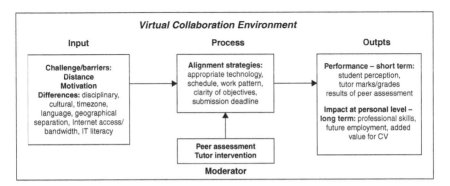

Figure 2.1 Model of online collaborative learning (after Soetanto *et al.*, 2012)

Distancing factors

Mismatched schedules

One element that all groups noted as an issue that created distance between the groups in Coventry and those in Ryerson was that of the overall schedule of work. Because of the timetable of activities, the Ryerson students began their activities before the Coventry students and ended before them too. Typical quotes by the groups concerning this were:

> We need to rush off at the beginning of the project in order to catch up with them. Assist them with their submission. The only problem is the submission date being different. We are not working at the same pace.
>
> (A4)

> The only problem is Ryerson's students start this way before us, so we have to meet their deadline . . . and our deadline not until next week. We have to do a lot of work a month before our deadline, so that put a lot of pressure to us.
>
> (A5)

Disciplinary difference

Another factor creating distance was the different disciplines involved. The groups at Ryerson were composed of architects; those at Coventry were civil engineers. One group noted that some of the issues they were encountering were no different from those that their colleagues engaged in face-to-face collaboration with Coventry-based architects were facing.

Task difference

Groups reported that the students at Ryerson had been set very different tasks:

> We had a problem at the start because the two universities did not have a common design brief. There was confusion over what they are doing and what we are doing . . . and the information we get and things like that, does not seem to tie in quite with everything else. So obviously, there seems to be differences between what they have been told by their lecturers, and what we have been told by our lecturers. So the two universities can sort a common design brief, and things like that, will be much helpful.
>
> (A2)

Different standards for different countries

Only one group had encountered any confusion due to different standards across the two countries, this was simply that "their standard for parking size is different from the UK" (A7).

Alignment factors

Familiarity with the technology

The students' familiarity with videoconferencing greatly aided this alignment. For many students use of Skype was very natural, to the extent that one group referred to meeting this way as face to face:

> Our communication was very effective cause we speak face to face via Skype once a week, and we e-mail each other regularly throughout the week. So it was quite effective.
>
> (A2)

Note that this was group A2, one of the groups whose collaboration was categorised as "unsuccessful". It seems that technical alignment, for this cohort of students, was a given and on its own did not enable effective collaboration. In fact, another group (A1) emphasised that the technical aspects did not, for them, constitute anything different from face-to-face practice, stating that "It's just a new thing, does not make a lot of difference to be honest".

The familiarity with Skype was also referred to by Group A6 (partially successful) in their response to issues about the technology they state "I do that every day anyway", meaning using Skype was not an unusual activity for them.

Meeting schedule

Another alignment factor commonly referred to by the groups was the creation of an aligned schedule of tasks and meetings. A typical schedule for meeting was a Skype call once a week, sometimes twice. This schedule was established at the beginning of the collaboration, and this basic schedule remained fixed. The more successful collaborations adopted the flexibility of additional meetings when required.

Flexibility was also important for the meeting structure and length. Adjusting the type of contact depending on the need and the project phase was also a factor employed by the successful groups.

> Normally last about a half hour. . . . it is only the first meeting was one and a half hours . . . two hours. . . . sorting out who does what task. Afterward, when everything becomes clear, meetings like a half hour maximum. Yesterday's meeting was 5 minutes.
>
> (A4)

Being able to make decisions about how to solve problems and make decisions is a highly useful practice in effective collaborations and is known as meta-decision making (Schuman, 1996; 135). Having a process by which the collaboration can come to a decision about length of meetings and appropriate schedules for these appears to have made a difference to these groups.

Recognition of difference and perceiving this as a strength of the collaboration

A factor that indicated a positive position on collaboration, and one adopted by the majority of groups, was that of recognising the complementary nature of the two skill sets and how the overall collaboration was stronger as a consequence of this.

> Exchange of experience. . . . they are better in designing layout, landscape, drawing . . . they are very good. I think they have more experience.
>
> (A3)

Structure of teams

Arranging meetings through Skype with the other end did not appear to be particularly problematic for the groups. More problematic was the difficulties in arranging for all of the near-side team to be able to attend the meeting. This was particularly difficult for A6 since they were part-time students. The groups found different solutions to the complexity of having two groups of four people all collaborating.

One solution was to structure the activities so that only one person worked with their opposite number to complete that task. Another solution was to appoint a single person, or two people, to represent each group at the meetings and have them co-ordinate the activity of the Coventry or Ryerson group. A third solution was to break down into more detail the tasks between the two groups so that individual members could work more independently.

Although adopting different strategies, it did appear that the groups tended to adopt some mechanism to streamline the collaborative activities, particularly during the synchronous activities, which required clustering four people around the desktop cameras at each end if all attended. Communication pathways among eight separate people asynchronously also amplified the complexity of the tasks to an extent that the students found too taxing.

Role of professional ethos and trust in distanced collaboration

Professionalism, the factor that has the biggest impact

The single factor that all those collaborations that were unsuccessful had in common and were different to those collaborations that were successful was the attitude to professional behaviour displayed (in their perception) by the team at the other end. Both of the unsuccessful groups (A2 and A3) had experienced problems with their experience of the work ethos of the team at the other end. Conversely, group A6 admitted that the fault lay on both sides, neither side always meeting their commitments on time, which they attributed to the mismatch in schedules. This was despite an excellent start as far as project management practice is concerned, in that they shared expectations and goals. But overall, their opinion of the group at the other end was positive, and this dominated.

Higher degrees of collaboration: meeting others' goals

One area in which even some of the successful groups struggled was in the attitude to different goals and tasks. As the students observed, a professional outlook requires people to make efforts to meet the objectives of other members of the organisation, not just their own. Groups tended not to work towards a common goal of completing the project, rather their own individual objectives. Only one of the groups achieved this higher level of collaboration. Their attitude was that it is an integral part of collaboration to see the whole team as one group with a shared goal:

> Oh, it is just report and stuff like that, any research that you can find. For example, I am surfing the net, and find some stuff on sustainability that might be of interest to them, and put it in, have a look at it, might be interesting. Just help them out, we are group at the end of the day.
>
> (A7)

Willingness to exchange information

This collaborative attitude towards the part of the team located at another institution also manifests itself in the exchange of information when it includes anticipating the needs of the other half of the group and passing on information which *could* be of use to them rather than only passing on the minimum that they request, in the close attention to the information given by the other group and the clarity of request for information.

In the interviews, the students detailed some of the experiences of exchanging information

> It was clear. . . . I give you an example over the task that I am doing . . . I do accessibility scheme, so I submitted the work, and they say no, we actually need some more information. They told me exactly what they need. They said in my report that there is going to be a ramp here . . . there is going to be a zebra cross here . . . they told me to include that in CAD drawing that show the two are connecting.
>
> (A4)

> Everything, even if we think it does not matter . . . we send it all to them and they sent it to us. Notes, sketches, CAD drawings, meeting minutes . . . everything! For instance, they sent us the drawings. For instance, one chap in our group noticed that the . . . pond won't work, got mention and got sorted. It was fairly quickly.
>
> (A5)

They gave us specification to read through. There are some blanks, and they wanted us to feed into it. They sent us an e-mail requesting a few things, and we had to send e-mail them back telling them it is a bit unrealistic at the moment. We do what we can, and we get over what we can. We gave them information for aspects but it was not report that they wanted from us.

(A6)

As can be seen, the first two quotes indicate that the students are freely exchanging information, but in the third, as the relationships between the two locations become more strained, the students' tendency is to limit the amount of information exchanged.

Consequences of lack of professionalism

The consequence of a lack of professional ethos had a greater impact due to the distanced nature of the collaboration, since the students had no recourse to alternative supportive forms of interaction. Although the distanced nature of the interaction is not a problem for those with a successful collaboration, it permits a range of additional concerns to arise when communication is taking place within an unsuccessful collaboration.

For example, for group A2, the fact that they only met for an hour and the work at the other side could not be constantly monitored became issues.

Q: Is collaborating over the internet as effective as face to face?

A: No, I don't think so. Although everyone can put on a face for an hour and speak effectively, you never know what they are doing, you do not know the quality of the work that they are producing until they give them to you. . . . You need to put a lot of trust in the other team to do what they say they are going to do, which we have been doing. But so far it seems that that trust is being misplaced.

(A2)

The increasing lack of trust was noted by the other team with an unsuccessful collaboration and was responded to by a similar desire to withhold information:

I am straight talking, but I am not going to be nasty to them. I am also very helpful, but if they start doing like this, we are going to remember after Xmas. We are going to be very reluctant to give them anything they want, without doing something for us. . . . we are struggling to get anything out of them. Next term, we are going to turn the table around . . .

(A3)

Again, the distanced nature of the collaboration exacerbated this breakdown in trust; because the other team could not be observed outside of the brief Skype meetings, this meant that there was an opportunity for doubts and suspicions to develop further. Ring and Van de Ven (1994; 93) define trust as faith in the moral integrity or goodwill of others, which is produced through interpersonal interactions that lead to social-psychological bonds of human norms, sentiments, and friendships (*Homans, 1962*) in dealing with uncertainty.

Trust between the organisations within a collaborative partnership is not a given, since the trust required by organisations is high within collaborations.

> The ability to presume that partners are acting responsibly towards and within a collaborative relationship is highly valued and likely to influence where commitment to collaborative activity is directed. To presume, however, that others are being open about their motives, interests, information, decision-making and changes in circumstance, at least as they affect the collaboration, requires heroic thresholds of trust, or acceptance of risk.
>
> (Cropper, 1996; 95)

Trust builds up effectively through incrementally increasing assignments and tasks set. By fulfilling these tasks healthy interorganisational relations (IORs) develop; if these tasks are not fulfilled, trust will diminish. This development of trust through repetitive and escalating commitment and execution of tasks can be seen through the comments of the groups involved in successful collaborations.

Consequences of displaying professional ethos

When the majority of the collaboration is successful and the perception of the team at one end is that the team at the other is taking a professional approach to their work, then other distancing factors become less problematic. For example: technological problems, such as Internet interruptions, are overlooked or adjusted to:

> **Q:** No interruptions?
> **A:** No, not really . . . only very odd occasions, it is just minor interruptions with Internet signals and things like that . . . very minor.
>
> (A5)

Also distancing factors such as language become insignificant or opportunities for bonding:

> There were few times, when they thought some of our sayings were funny and we did not know what they were saying because it was not what we say.
>
> **Q:** Is it a problem?
> **A:** Not really, it is just funny.
>
> (A5)

Even mismatches in task briefs were more easily overcome, or overcome with more confidence, if both groups of students behaved with a professional attitude:

> **Q:** Is there any instance when you felt doubt about the work?
> **A:** In the past a couple of weeks, yes, because it was not very clear about what we were needing to be doing. But the collaboration work between our work and the Canadians were fine.
>
> (A5)

As were mismatches in schedules:

> No more meeting as they have done their part, and now we need to do our part. They have other coursework to do . . . not until January. Once they submitted and uploaded, and let us know, what do you think? Thanks for working with us. It has been great and see you in January.
>
> (A7)

> We have different tasks next term, but in terms of what we have achieved so far. It has been really good. At the moment, we are working to submit our. . . . and they have submitted theirs. You got a gap that you can talk to them . . . they have exam or holiday.
>
> (A4)

For students in less successful collaborations, these were seen as the elements that were causing barriers in effective communication and were seen as major hindrances in establishing good working practices. For the students in the groups quoted here, these are seen as minor distractions, having little or no impact on their ability to work together.

As noted above, the main factor in the development of trust is to complete tasks on time, but trust is also tied into presentation of efficiency within the videoconference.

> They might have problems over there, but when they have a meeting with us, they were on the ball. . . . they were very professional and very competent when they had a meeting with us.
>
> (A5)

Although, as described in the section on the consequences of the lack of professionalism, putting on a show of efficiency for an hour in Skype is soon seen through when tasks are not delivered on time.

Development or diminution of trust through performance

These two cycles, one of increasing trust and commitment and one of diminishing trust and commitment to the collaboration, can be summarised in Figures 2.2 and 2.3.

Figure 2.2 Cycle of decreasing collaboration in virtual teamworking

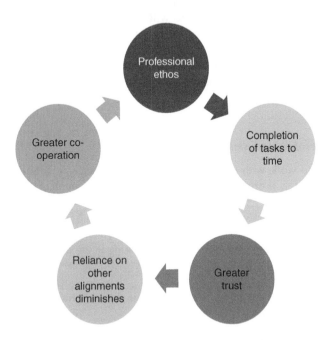

Figure 2.3 Cycle of increasing collaboration in virtual teamworking

What emerges from these two sets of experiences of the students are two positive reinforcement loops. One of these derives from a professional ethos between the students. Tasks are completed to time, trust increases and a willingness to co-operate develops, and this results in a tendency to help others to meet their goals, and therefore tasks are more likely to be completed. In this cycle, issues with the technology, scheduling, and cultural differences are seen as only minor distractions. Conversely, failure to meet deadlines leads to lack of trust, the two parts of the team at each end withdraw from open communication and completing tasks becomes less likely. This is a cycle that would exist in a co-located collaboration. However, in a distanced collaboration the additional set of barriers introduced by the technology exacerbates this problem. The students see the technology and scheduling difficulties as difficult barriers to progress through and are focused on as problematic areas. Usual strategies for repairing conflicts that occur in face-to-face situations are not available to them, and the trust further diminishes.

Impact factors

Within the input–process–output model adopted for the analysis, the outputs for the participants were either in terms of grades improved or in terms of impact on their own professional development. Taking part in this activity was not seen as likely to enhance grades. For the Coventry students, collaborating online with architecture students in Toronto was seen as a more complex and risky task than collaborating offline with architecture students in Coventry. Taking part in the online collaboration was therefore seen to be likely to have no or even an adverse effect on their marks. The predominant rationale for taking part, cited by most of the groups, was that they thought it would be interesting. The three other reasons mentioned by the groups are listed in what follows.

Acquiring professional skills

The ability to work with other nationalities at a distance was thought to be an intrinsic part of their professional career as construction engineers, and the online collaborative activity would provide a work-like experience that would prepare them for this sort of activity. Typical statements were:

> It gives you a simulated experience of working in the industry.

> (A3)

> It is experience of working with people you do not know, from different educational background.

> (A7)

How to work with other people who have other idea and other . . . they have been taught in different way . . . if we worked with architects in this university, we knew that they were taught in certain way. It is just the experience to work with international programme.

(A5)

Bringing it all together and making it work is something that you have to do in the future in reality.

(A7)

Working with people from different educational background and stuff. Just like, prepare us for the future, if you work with people from different countries, help you to engage people better.

(A7)

An interesting correlation was that those groups that tended to have the stronger collaborations also had the clearest ideas of the value of the work in their future careers.

Of interest to potential employers

That it would look good on their CVs was also an attractive prospect for most of the groups. The experience was particularly prized because it was thought to be a rare one, and so it would give them a greater competitive advantage when looking for work.

Good experience, not many people get to do this.

(A5)

The thing with this is when you go to apply for jobs. They see that you've worked internationally, already, without leaving the country.

(A5)

It is just to see how people can be like across different country . . . you know. . . . also, main reason is we can put in our CV.

(A7)

It is good on CV as well.

(A1)

Again these statements came from the more successful collaborations.

Easier than face to face

Finally, one group (but only one) commented that actually conducting this collaboration at a distance rather than with another local group of architects was preferable because it was actually *easier.*

> It is probably easier. Some of the groups I've spoken to that are working with other Coventry students, they seem to have more problems than we do.
>
> (A6)

Although the technology adds barriers to the communication and the ease with which communication can happen, it also presents aspects that make collaboration easier. Although this was not a theme that was explored in this first project, the second project explored this more, and the advantages of distributed rather than co-located collaborations are discussed in the following chapter.

Conclusions regarding the undergraduate experience

The responses from this particular cohort of students indicate the following:

The mismatching in schedules at the two sites, the perceived difference in the tasks set and the fact that the team was a virtual one did not have an impact on the teams that were successfully collaborating

Those teams that successfully collaborated managed to effectively work around these issues with little problem.

The virtual nature of the collaboration only had a negative impact when the collaboration was not a successful one

The distance factors, including the contact being so limited and the fact that virtual co-workers could not be observed when out of meeting led to additional doubt and suspicion about the commitment and efficiency to the work. Some reported a positive impact.

The single greatest factor in supporting successful collaborations was the professional ethos of the groups and the consequent building of trust

On-time completion of tasks and effective performance in meetings built trust and increased the collaborative nature of the teamworking. Attitude to collaboration was therefore the defining variable in whether a collaboration was effective. As all groups chose the same technologies and some were successful and some were not, the technology does not appear to be a defining variable.

Only the very most collaborative team aimed to support the objectives of the other team if this lay outside of their own task briefs

Students, on the whole, only did the work they were committed to do. Only one of the teams aimed to support their colleagues at the other end through offering additional help and support.

Effective collaborative teams also display meta-decision-making processes and openly and frequently share information

All teams had a regular schedule of meetings. However, only the most collaborative of teams had the flexibility to add additional meetings when these were required by the project phase or individual tasks and would vary the style and length of meetings to meet these needs. Also only the most effective collaborators shared all information "just in case" rather than keeping information to the level of that demanded.

The virtual teamworking mode was not a barrier to teamworking

Students felt that virtual teamworking was "just something new", that is something novel, but not so outside of their experience that it represented a challenge. Many already used Skype frequently for personal use. One group felt that it was probably easier than face to face for scheduling meetings. All found it an interesting mode of working.

Difficulties in forming a single unit led to strategies for structuring the group

All the groups found problems with holding a single meeting with all the participants from Ryerson and Coventry in attendance, and most did not try. Three strategies were developed to avoid this; one was to have a single representative or two representatives of each group at the meetings who could then act to funnel information to and from the rest of the group. Another was to divide the whole group into pairs of students working in twos virtually. A third strategy was to break down the tasks into individual tasks so that the students were working independently rather than collaboratively.

This third strategy, however, undermines the very idea of team collaboration. Commonly and intuitively, the students immediately split their tasks into several chunks of work which would be shared with the individual members. In future, it has been suggested, the coursework briefing could be improved by focusing on higher (i.e. reciprocal) level of task interdependency.

Students felt that the activity would have a positive impact on their employability

All students felt that this activity reflected the nature of international collaboration in the construction industry, and this experience would be highly prized by potential employers. All felt too that this gave them a greater competitive advantage in the jobs market since few other students would have this opportunity.

Virtual teamworking with postgraduates in construction engineering

Simultaneously with the virtual teamwork module being conducted with undergraduate groups, a separate task was conducted as part of a postgraduate module. The students on the course were a mixture of full-time students and students who were full-time and part-time workers in the civil and construction engineering industry. These latter groups have therefore awareness of workplace culture and communication methods. For this task, the groups were asked to adopt virtual teamworking methods for part of the collaboration over a course of two or three meetings. For three of the groups, who could have met face to face for all the meetings due to being all co-located throughout their collaboration, virtual teamworking was therefore an academic exercise, purely to fulfil the requirements of the coursework. These are referred to as the co-located groups; those that needed to meet at a distance on occasion are referred to as mixed-mode groups.

As the students all met frequently face to face it was not possible to draw links between their use of technology and a successful collaboration. Instead the groups have been identified as whether they displayed an overall positive position on the ability of videoconferencing to convey communication factors (such as body language and facial expressions) or a predominantly negative one, determined by the number of positive statements made compared to the number of negative ones. The same framework of distancing factors, alignment factors and impact are used for grouping categories.

The data were collected from submissions made as part of their coursework, from the section in which they were asked to reflect on the nature of the virtual communication. Not all the groups chose to reflect on their own personal experience but instead made comments based on the literature regarding face-to-face

Table 2.1 Summary of the responses to the virtual teamworking task

Group	Entirely co-located or mixed mode	Predominantly reflective report or predominantly a literature review	Predominantly positive reaction to videoconferencing or predominantly negative	Main mode of communication in meetings
C1	Mixed	Reflective	Positive	video
C2	Co-located	Reflective	Negative	video
C3	Mixed	Reflective	Positive	video
C4	Mixed	Literature review	Negative	audio
C5	Co-located	Reflective	Negative	video
C6	Mixed	Reflective	Mixed	video
C7	Co-located	Literature Review	Negative	text

and virtual communication. In some cases, therefore, data were limited. Students were not available to ascertain the reason for this omission, but staff and students in previous research projects within the subject discipline have raised issues concerning the value of personal reflection on experience as credible content for submitted work. This may have led the students to overlook their own experiences as sources and choose to report on the literature instead.

The various categories the groups fall into are summarised in Table 2.1.

Differences in the experience of postgraduates when compared to undergraduates

Distancing: IT and information literacy

Some of the groups appeared to struggle with some of the IT skills required to implement the technology. For example, Group C5, whose members had rather a mixed experience of using the technology, reported the following experience of Dropbox (a relatively straightforward and standard program):

> Although the software was easy to download we found it difficult to use the software, especially when trying to open other people's work on your own laptops. The instructions were very technical and difficult to follow, therefore we decided not to use the software throughout our project.

Others had little experience with Skype, again a very common program. For example group C2 stated that although "Some of group had used this form of medium before . . . some had not", group C3 stated that "one member of the team had not used Skype before, which delayed the first conference as they had to subscribe and get acquainted with Skype". As a response this also indicates a sufficiently poor digital literacy (or motivation to engage) to not ensure the technology was working before the meeting. Others chose not to attempt to use new technologies simply because they were new:

> The virtual rooms were not utilised as they were unfamiliar and an initial face-to-face meeting resolved to conclude that the time frame for the project was too short for all group members to become sufficiently familiar with the technology.
> (C7)

This lack of experience with the technology was also in evidence in one group's usage of email. Because of their difficulty in learning how to use Dropbox, group C5 resorted to sharing documents via email, which led to issues of version control.

> On several occasions when replying to an email the sender did not click "reply all" so many emails and information was missed between members in the group which sometimes led to some confusion on "version control" of documentation.

The use of the quotes around "version control" is also indicative that this is not a term they use frequently and see as some form of jargon.

Email checking was also not a built-in part of the daily routines of some of the members in groups, again indicating a lack of experience with fully engaging with technology:

> People check their emails at different times in the day, so some time delay to questions was common, on one occasion a member of the group did not check their email for a number of days and missed several emails and responses therefore was out of the loop for a time.
>
> (C5)

Use of email for circulating documents inevitably leads to different versions being produced unless a strict timetable is adhered to concerning turn-taking in producing an updated document. The group were insufficiently aware of this practice to avoid running into problems:

> Sometimes more than one member of the group made an adjustment to the document in the attachment of the email, resulting in several versions of the same document being circulated, resulting in confusion of important information potentially being missed.

Other groups also were not used to regularly checking emails. For example group C4 stated that "The problem we had in using communication through emailing was the time each individual took to reply." Group C1 commented in their suggestions for improvement in team working that

> Group members should schedule times when they will check their emails throughout the day. This will give certainty to senders that their messages will be viewed in a reasonable amount of time.

This indicates that this is not common practice at the moment.

Distancing: perceived limitations of videoconferencing as a medium

Most of the groups also referred to the difficulties they had with the essential nature of videoconferencing as a medium for conducting meetings, in contrast with the undergraduate groups, who in general were familiar with it as a platform. Group C1, which had a generally positive experience of the use of technology, noted that

> web-conferencing meant that meetings progressed quite slowly, and this was quite infuriating at times. Furthermore, more can be discussed in the same amount of time, and it can also be discussed more clearly, when meeting face-to-face.

This slowed interaction was noted by several other groups, observing that "the meeting went well but lasted for longer than what would have been taken if the meeting was conducted 'round a table'". Although noting it, Group C6 did not find any essential problem with this slowed nature and simply modified their meeting structure to compensate:

> Both meetings did however progress differently, during the face-to-face meeting everybody got more involved, whereas during the Skype meeting there was one person who naturally chaired the meeting and steered it in the right direction. For this reason the Skype meetings did carry on for longer as there was a need to ask everybody if they agreed rather than seeing a nod or people saying yes when they were looked at.

A factor that more groups found difficult to deal with was the reduction in visual cues from other participants, for example facial gestures and body language. In fact, some groups claimed to perceive an *absence* in visual cues, even though they were communicating via video and so therefore could, presumably, see each other.

> Aspects of communication, such as body language, facial expressions, etc, can also be lost when web-conferencing.
>
> (C1)

> During the meeting it is hard to tell if everyone is paying full attention and understands what is being discussed, this is because you cannot see their facial expression unlike face to face. This led to periodically asking if people understood and agreed. Not being able to see each other meant that you did not know if someone else was about to start talking, this meant that in order to avoid two people talking at once (people) did not contribute as much as they would normally.
>
> (C6)

> Due to the 'barrier' of the 'interaction' on the screen, the communication felt devoid of personality and purpose.
>
> (C2)

> There was no sense of relationship gained, which is key to understanding of different groups and successful communication.
>
> (C2)

> Skype is brilliant but does not replace or provides the user with any sense of personality, empathy or trust.
>
> (C2)

Group C7, when asked about improvements to the technology that would make videoconferencing effective stated that

> Advances in technology could result in 3D holographic representations of each member thus increasing the ability to read body language and

increasing personability between group members. Even abstract ideas may be discussable.

In other words, it is only with technology advancing to the point at which the presence of the people at the other end matches that of the physical world that the interaction feels personable and abstract ideas can be discussed, rejecting the idea that these aspects can be quite easily communicated via videoconferencing.

The issue of this perception of lack of facial expression and body language is discussed further in the Conclusions section.

Two groups had members who faced issues with members having English as a second language. Group C7 found that this was exacerbated when in a videoconference:

> One of our group members is foreign and English is not his first language. During some of our audio conferences he expressed that it was difficult for him to keep up due to the deterioration in the sound quality. We resolved to try to speak more clearly and back what we said by text.

However, group C1 found that videoconferencing was actually supportive of language difficulties:

> Another advantage webconferencing has over face-to-face meetings is that a log of what was discussed is automatically generated, and this is useful for reviewing purposes. To have any such log in a face-to-face meeting would require someone to take minutes of meetings, and therefore be distracted from the meeting. This ability to review discussion was particularly useful for users whose first language isn't English. And it also helped to avoid any misinterpretation of certain words or phrases.

Alignment: adapting meetings to project phases and requirements

One area in which the groups did show flexibility was in the variation of the structure of the meetings. For example group C5 conducted the meetings thus:

> We used the first meeting to review the Sustainability Checklist and to identify and agree the issues relating to the project. The second meeting was used to formally check the progress of work and to agree deadlines for printing and for submission. We found that the meetings were very beneficial because they allowed for fluent discussions in real time between all members of the group. Decisions were made instantly and action plans for the future were agreed.

Meetings were structured according to the tasks at hand, with reference back to the project plan or brief.

> The conference in essence used the project brief as a template for setting the agenda, thus the discussions covered the seven areas that were required

for the final report. During the lecture week the seven areas were distributed between the team members.

(C3)

In actual structure and achievement, the groups all were satisfied with the meetings, even though the process was difficult for some, a typical example being:

Both the face-to-face and Skype meetings were successful because we managed to discuss and come to an agreement on everything we needed to.

(C6)

Impact factors: more limited rationale for engagement

As the postgraduates were people who already worked in the construction industry, the value of the acquisition of skills for distanced collaboration was not a driving factor in motivating them to take part. In fact, the prevailing opinion was that this was not a skill they had needed to acquire at this point in their careers, and so there was no need to acquire it. However, for some groups, distanced collaboration was an advantage in completing their assignments, for two reasons:

Of the four groups that had a generally negative set of comments to make about the experience, three of these were co-located and so had no actual reason to meet via videoconference. One of these (group C2) stated that "it worked as much as it served its purpose of completing an element of the coursework", the other stated that:

Although we can see the benefits of web-conferencing it was a purely academic exercise as all four members lived in close proximity to each other, therefore it was easier to meet face-to-face to make decisions.

All of the groups recognised the value of videoconferencing for conducting meetings with people in other countries. For example, group C5 stated that

We all agreed that there is definitely a place for web-conferencing in the modern construction industry. However, it must be treated with the same professionalism and organisation as a face-to-face meeting. "Web conferencing does not make meetings worse, but it can amplify the problem" (*Tom Austin, 2006*). If done correctly then we think that webconferencing can be a very effective tool but it cannot replace meeting face-to-face.

However, only three of the groups actually had to work at a distance as an essential part of the task, due to members of the group returning to homes that were distant from Coventry.

For us Skype was a must because one team member lives in Manchester and another went home to India for the Christmas vacation.

(C6)

Felt that the video conferencing was good in comparison to face to face meetings as it made it possible for all the members to have visual communication from different locations within the UK and overseas (i.e. Nigeria). This was useful as due to the fact that some of the (group C3) members worked full time while others were abroad, Skype made it possible to have visual communication at an amicably convenient time for all the members.

(C3)

Web-conferencing offers the advantage of connecting people over long distances. This was useful for our first meeting as one of our group members was away in India. It was also useful for our second meeting as it meant another group member did not have to commute to a meeting location, and therefore saved themselves time and money.

(C1)

These three groups were also the three that had the highest proportion of positive comments to make about the exercise, two being overall positive and one being evenly distributed between positive and negative. This may mean that, although intellectually groups were able to perceive the value of videoconferencing, it had to actually have an authentic role in their collaboration to be fully appreciated.

Conclusions regarding the postgraduate experience

Perceived absence of body language

The reason for the perceived absence in body language is difficult to identify. A fuller description and response to this experience was given by one student, who took the reflective piece submitted as part of the submitted work as an opportunity to put forward his ideological opposition to remote working.

Perhaps this is because the users were not familiar with the "protocol" of the medium itself, but something was missing. There was no sense of relationship gained, which is key to understanding of different groups and successful communication. . . . Within the same country, to make the sacrifice of an "extended" "working" day than is usual is a small price to pay, to not only understand but also appreciate the gestures and "human" mannerisms what are simply "lost" through web based medium. The argument of almost being forced to use these medium because "they" are "available" almost makes a mockery of the simple, face-to-face interaction humanity has used for millions of years.
      . . . that the medium is not able to fulfil is the interaction and instinctive behaviour, the human species are able to pick up from being in the company of others, which in itself, generates warmth, comfort and conflict which are all healthy products of human nature which manifests itself into trust; something the electronic age will never be able to achieve, despite the advances we have all experienced.

> Skype is brilliant but does not replace or provides the user with any sense of personality, empathy or trust, if I need to adapt my personality by the use of "smileys", detailing facial expressions suggestive of "moods", I would rather have less sleep, spend more money and travel by car or train and meet face-to-face, because sincerity can never be replaced by artificial intelligence.

This was of obviously deep concern to this group and constituted a significant proportion of the work submitted as their assignment and reveals much about the position of the group of people who had a negative experience of the technology.

One explanation is that this represents a general antipathy towards technology, and in fact another comment by Group C2 was that

> All four members were able to "join" the meeting. From a personal point of view and one shared by the group, the time leading up to the event was full of scepticism.

However, group C1, who identified many positive aspects of the virtual team-working, also took issue with the perceived lack of body language and facial cues.

The problem seems to be an absence in the ability to perceive the person at the other end and for any interaction taking place with an image to be inauthentic. For example, in the earlier quote group C2 uses quotes around the word "join", indicating that attending virtually is not really joining a meeting. Group C2 also made the following statement:

> The act of talking at, but more importantly talking into the screen, did not give the same sense of passion to convey the information one would be able to in person.

The use of language recalls that of the students observed by Bayne (2004), recounted in the introductory chapter, who could only perceive online interaction as inauthentic.

It may be that there is a rejection of newer technologies that is prevalent within the workplace from which these students come, which views any form of communication that is not face to face as essentially lacking in the required level of authenticity.

Comparisons of the undergraduate and postgraduate experiences

Differences between the two cohorts

The postgraduate groups displayed many differences from those in the undergraduate cohort. The distancing factors that they experienced that the undergraduates did not were

- Difficulties due to insufficient IT literacy
- Hardware issues
- Perceiving problems in communication due to not reading body language

These issues tend to indicate that, in general, the postgraduate cohort had an unfamiliarity with many of the technologies and an experience of unease when dealing with the videoconferencing aspects. Many groups reported members with poor work practices in dealing with email and version control of documents. Some groups admitted to a general scepticism about the use of the videoconferencing technology. Where the postgraduates displayed fewer distancing factors was in the actual collaborations. None of the groups discussed failures of members to provide information or meet deadlines.

Similarities between the groups were that both employed some similar techniques to promote alignment. These were closely defining tasks and allocating these and identifying and agreeing on shared technologies for communication. Both cohorts also ran highly structured meetings with meta-decision making processes, that is, they would flexibly decide on how to reach decisions at different stages in the project. In short, the issues displayed by the students in the postgraduate cohort were limited to the use of the technology not to generic project management skills.

Potential causes of the differences between the cohorts

One difference, and perhaps the major defining difference, was that the activities were assigned to all of the students in the postgraduate cohort, whereas the undergraduates could opt for either entirely face-to-face collaborations or to collaborate with a team in Canada. This would mean that those undergraduate students who had particular aversion to using technology would not have taken part in the activity.

A second difference is the age of the cohort. The postgraduate group involved many students from industry who were of a slightly older age group. These people might not use communicating technologies to the same degree as the traditional students of the undergraduate cohort.

A previous study, conducted by Childs and Espinoza-Ramos (2008), identified a range of different technologies employed personally by students and found that international students were more adept at using distanced synchronous technologies such as Skype and MSN Messenger. In the postgraduate cohort, those collaborations which involved people who were located overseas during the collaboration were more positive about the experience of working with technology. This could be because of them being more likely to use distanced communication in their personal lives (as in the findings of Childs and Espinoza-Ramos) but is also likely to be because these groups had a necessary rationale to use the technology, as it was the only way for them to communicate.

Lombardi (2007; 4) when describing the role that authentic learning can play in students' learning, states that:

> Educational researchers have found that students involved in authentic learning are motivated to persevere despite initial disorientation or frustration, as long as the exercise simulates what really counts – the social structure and

culture that gives the discipline its meaning and relevance. The learning event essentially encourages students to compare their personal interests with those of a working disciplinary community:

For the postgraduates with experience of working in the construction industry, the distanced collaborations may have been of value when it was the only way to continue the task, but for those that did not, it had a doubly inauthentic purpose. First, it was not required to complete the collaboration (merely because it was requested as an assignment), but second, it was not providing them with the lived experience of the working practices of the construction industry, because in their experience, the construction industry does not use distanced collaborations. For the undergraduates, it was authentic because their team was split between Canada and the UK, and the experiences the task was providing for them was (they presumed) of relevance to their working careers.

In essence then the differences between the cohorts and their experiences of working online arise from four different causes:

1   The undergraduates were self-selected and the postgraduates were not, which led to some of the postgraduates belonging to those unable to fully experience online interaction.
2   The undergraduates could not be co-located, which meant the interactions had to be online, and so the videoconferencing was not simply there to be assessed but was an integral part of the activity. Some of the postgraduates had no barriers to meeting face to face, so imposing distanced working was an unnecessary barrier.
3   The postgraduates had experience of construction engineering practices, and for none of those up to that point had this involved distanced collaborations. For them, therefore, the task did not convey relevance for that discipline. For the undergraduates, who presumed that distanced working was an integral part of working practices in the industry, distributed teamworking was not only relevant, but practising it as part of their undergraduate studies provided them with a competitive advantage in the job market.
4   The undergraduates were younger and therefore less resistant to modern methods of working which incorporate more technology.

As the industry changes to encompass more distributed teamworking, this could well lead to younger workers having an advantage, as they have a skill set – and mindset – more appropriate to working and collaborating online.

References

Bayne, S. (2008). Uncanny Spaces for Higher Education: Teaching and Learning in Virtual Worlds. *ALT-J*, 16(3), 197–205.

Childs, M. (2010). Analysis and Description of Education Employing Technological Platforms: Terminology, Features and Models, in Clouder, L., and Bromage, A. (eds.),

Interprofessional E-Learning and Collaborative Work: Practices and Technologies (pp. 46–60). Hershey, PA: IGI-Global.

Childs, M., and Espinoza-Ramos, R.M. (2008). Students Blending Learning User Preferences: Matching Student Choices to Institutional Provision, in Hodgson, V., Jones, C., Kargidis, T., McConnell, D., Retalis, S., Stamatis, D., and Zenios, M. (eds.), *Proceedings of the Sixth International Conference on Networked Learning*, 492–499, 5th–6th May, 2008, Halkidiki, Greece. www.networkedlearningconference.org.uk/past/nlc2008/abstracts/PDFs/Childs_492–499.pdf.

Cropper, S. (1996). Collaborative Action as a Model of Conduct, in Huxham, C. (ed.), *Creating Collaborative Advantage* (pp. 80–100). New York: Sage Publications.

Heeter, C. (1995). Communication Research on Consumer VR, in Biocca, F., and Levy, M.R. (eds.), *Communication in the Age of Virtual Reality* (pp. 191–218). Hillsdale, NJ: Lawrence Erlbaum Associates.

Lombardi, M.M. (2007). *Authentic Learning for the 21st Century: An Overview*. Louisville, KY: Educause Learning Initiative.

Moore, M.G. (1993). Theory of Transactional Distance, in Keegan D. (ed.), *Theoretical Principles of Distance Education* (pp. 22–38). London: Routledge.

Ring, P.S., and Van de Ven, A.H. (1994). Developmental Processes of Cooperative Interorganizational Relationships. *Academy of Management Review*, 19(1), 90–118.

Schuman, S.P. (1996). The Role of Facilitation in Collaborative Groups, in Huxham, C. (ed.), *Creating Collaborative Advantage* (pp. 126–140). New York: Sage Publications.

Soetanto, R., Childs, M., Poh, P., Austin, S., and Hao, J. (2012). Global Multidisciplinary Learning in Construction Education: Lessons from Virtual Collaboration of Building Design Teams. *Civil Engineering Dimension*. 14(3), December 2012 (Special Edition), 7–18.

3 Experiences of online collaborative design in asynchronous and synchronous working

Findings of the BIM-Hub project

Mark Childs

Introduction

Following from the Learning to Create a Better Built Environment project, a second implementation of online collaborative working was created as a three-way collaboration among Loughborough and Coventry Universities in the UK and Ryerson University in Canada, the full details of which are described in Chapter 1. Altogether seven groups took part, comprising (at least) two students from each university. The findings of the learners' experience of this project, presented in this chapter, are based on qualitative data gathered from 18 months of data collection from students and staff from Loughborough, Coventry and Ryerson universities.

This second iteration of the collaborative activity with the students built on and refined the previous study. It therefore forms a second cycle of an action research process, discussed in detail in Chapter 5. A different structure was adopted for analysis of the data, and more data were gathered. However, some of the findings mirrored those of the previous activity. To avoid duplication, the factors presented here are those that enhance and extend the findings of the previous study.

Data were gathered from three data collection methods.

- Focus groups with the three universities. Two of these were structured conversations with students; those with Loughborough were structured conversations with additional feedback in the form of flipcharts. Focus groups were conducted at the three institutions, with Ryerson University via GoToMeeting and face to face at Coventry University and Loughborough University. At Coventry and Loughborough, students sat in the groups in which they worked. In the first semester, Coventry students were asked a set of questions and were engaged in a conversation as separate groups (identified by the letters CA1 to CE1). The Loughborough students were asked to fill in flipcharts based on a set of questions, then fed back their answers as a class (L1). Ryerson had a single focus group (R1). In the second semester, Loughborough and Ryerson students were interviewed in a single focus group, and these are referred to as L2 and R2; Coventry students were interviewed in their groups (CA2 to CE2).

- Personal reflections from students. Students from all three universities were assigned the task of writing personal reflections. A random selection of these from Loughborough and Coventry were selected for analysis. All of the Ryerson assignments were analysed, as these were fewer in number.
- Analysis of video recordings. Students selected one video recording of their GoToMeeting meetings for submission as a document of their process. These were viewed and analysed, and observations led to guidance notes for online synchronous collaboration.

These three methods enabled triangulation of data of the experience of online collaboration and indicated that all three methods were producing similar sets of data. The responses were organised according to a different structure than the previous analysis. Throughout discussions with other members of the project team, the principle emerged that the factors influencing the effectiveness of the engagement could be grouped into a series of sets, each one dependent on acquiring or putting in place factors from the layer below it. These sets of factors are shown in Figure 3.1.

Levels 1 and **2** will vary according to the subject disciplines. For the BIM-Hub project this was specifically in the fields of construction engineering and

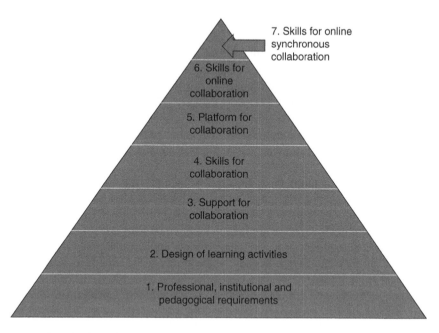

Figure 3.1 The dependent factors in developing effective online collaborative working

The author wishes to acknowledge the contribution of Zulfikar Adamu to the model

architecture. The remaining levels relate to generic skills that apply to any collaborative design activity that students may undertake.

Levels 3 and **4** will be relevant for *any* collaborative project, whether offline or online. Because all of the face-to-face collaborative projects that our students had experience of taking place within their own institution, all of these skills presume intra-organisational projects – though they may be inter-disciplinary.

Levels 5 and **6** refer specifically to those projects that take place online, when the teams are working collaboratively in a virtual environment. This is also the point at which many of the skills at working on inter-organisational and international collaborations can be acquired.

Level 7 is specifically for the additional skills required when activities take place synchronously, usually during videoconferencing. Some skills appropriate only to international collaboration can also be acquired at this level.

As this chapter focuses directly on the learner experience, the analysis begins at level 2. Level 1 is not a level in which the students have direct involvement, and as the contributing factors relate to the professional and institutional context for the activities, these have already been explored earlier in this book (Chapter 1). Although not responsible for the design of level 2, the students did have feedback on this level and made valuable observations regarding their needs and suggestions for an effective learner experience. The analysis therefore begins with level 2.

Learners' experience of the support for their collaborative working

At level 2, the way in which the learning activities are designed includes both the sequencing and structure of the tasks, the design briefs and the manner of the assessment, all of which have elements that are additionally problematic when moving into a collaborative online environment. The issues identified by the students make valuable contributions to how educators need to prepare and organise the learning environment for their students in these forms of study.

Lack of consistency in brief

As with the previous study, there was a lack of consistency in the brief given to the students in the different universities; communication between staff and students within the separate institutions was thought to be very efficient, however.

However, in this study students offered a solution: to create a single portal where all information could be shared. This is eminently simple and practical, since all tutors and students would be viewing the same materials and so therefore could observe where messages were going out that were different to their perceptions and could be challenged. This would also provide a common repository for all materials relevant to the students. This was also suggested as a resolution to the submission problems; everyone would know the date of submissions and be able to have a single working place to submit them.

> What we are talking about is a discussion board, when one guy asks a question, we all find out the answer. So that it would come out to everyone and everyone is in the same boat.

(L1)

Issues with assessment

One of the reasons suspected for the difference in motivation among the three sites was the differences in weighting of the module. The Loughborough students felt that they were less motivated to contribute because the final marks were worth a smaller fraction of their overall marks for the year.

> Another thing that affected that was the actual weighting for this module between the universities was different. I think this is work a 1/3 or 1/6 of their entire year while for us it is only 1/12. That, in terms of work, and their view on people's input affected that massively. And I think for Canada their commentary was worth 20 comments.

Students felt that the peer-assessment part of the process was inherently unfair, since students could unfairly bias the marks by voting for the students at their own location and voting up those at other locations if there had been conflict. The trust issues that can arise within collaborations could therefore manifest themselves in the grades they receive.

> Peer assessment bias: it's difficult because we, in our group, said to Canada will give everyone 100% because everyone put as much as they can, no matter if anyone let down, we gave full marks. But we definitely didn't receive a thing from Canada or what not. So remove the PA system, that's my advice because it's easy for someone to get on the computer and because they didn't like him they marked him down.

A proposed solution for this was for peer assessment to be a group exercise in which all the participants in the group jointly identify a mark each person in the group deserves.

> The group should come together and be like – he deserves 90, he deserves 100%, and amongst themselves they decide. And if they have any problems then they go to the lecturers and say to them "we've got issues; we'd like to give this person 100%, this one 105%, etc."

An alternative means to judge each individual's contribution was to track their contribution to the shared drive where materials were stored. Although this takes away from the concept of a shared group submission, resentment about peer assessment from some students was strong enough for them to consider this as an option.

> When you hand in your work you hand in what group X has done. It's one group's work. But it would be better if you could get each place to upload the work they've done so you can see how much each person has contributed and that will take away from the peer assessment. So lecturers can see that 90% of the work was done by these 3 and 10% by this person.

Another suggested resolution to the problem of peer assessment was to "Remove PA (peer assessment) – any issues go directly through lecturers", indicating that their view of peer assessment was as a punitive resolution to issues of non-performance rather than to encourage reflection and support peer learning.

Issues with the structuring of the assignment

A particular problem for Ryerson was that the structure of the task meant that a lot of the initial work fell to Ryerson students, as this required design elements. This was not seen to be due to a lack of commitment from the other universities, simply that this fell to the Ryerson students as it was their specialism.

> If there are issues between the teams it would be I think that because we're focusing on design it's heavy on what Ryerson's having to do right now. Loughborough and Coventry are supportive of the project. It's just that's what's needed for the project at the start there's a lot of design work that needs to be done as opposed to structuring, costs, It just takes a lot more effort to do a lot of the work that the groups are expecting Ryerson to do on the design point of view.
>
> (R1)

Observations on students' comments

Evidently online collaboration becomes an issue for staff members as well as students. Regular meetings and ensuring all decisions are shared by all the institutions are as essential for planning the shared activity as they are for carrying it out. However, these are skills that are as rarely embedded in the normal practice of educators as they are in the experience of students. Some initial awareness raising amongst staff of some of the core skills and protocols required in online collaboration are therefore necessary. Regular meetings and storing all materials centrally are essential. If last-minute revisions may be needed in the planning of activities in putting together a module, these cannot be put into place independently, as this has an effect on the other institutions. A shared portal, so that the briefs cannot diverge through these last-minute changes, and ensuring all students at all institutions receive the same brief is a practical way to ensure this does not happen.

Peer assessment is regularly criticised by students; it is a means of assessment that creates discomfort amongst many and is seen as one that can be easily gamed by other students. Part of the process of integrating peer assessment within tasks is to explain the purpose of it. This is to enable students to learn to reflect on their own and others' work and to judge where improvement may be possible and thereby develop the criticality necessary for personal and professional development. Although it also has a degree of providing external motivation to perform well and to add a level of discernment between marks (enabling those who contribute less to a combined effort receiving fewer marks), it is not primarily punitive in nature, although some of the students have perceived it as such.

In addition, despite different grading systems at the differing universities, it is important that the marking is as equitable as possible across the collaboration. Enabling students to see the marking criteria from their colleagues' institutions is necessary for transparency and in understanding the motivations of the other students. Placing everything within a single portal again addresses this issue.

Timing and scheduling of tasks and submission dates may be the most difficult of the aspects raised by the students. Semesters begin and end at different times, take breaks at different times and require submission at different times. Workloads also need to be balanced against other demands in the students' courses, which becomes much more complex the more institutions are part of the consortium. There is usually no perfect solution that avoids all of the problems scheduling creates, but being aware of where some of the issues may arise and developing strategies to mitigate these needs to be part of the planning stage. What will one part of the group do before the others begin? How does this affect allocation of marks? How can the task design take into account different submission dates at different institutions? All of these types of questions need to be addressed. However, these sorts of issues also occur in the working environment, so to some extent, learning to adapt to these is part of the learning experience for students and adds to the authenticity of the task. Foregrounding and explaining the value of learning to work within these constraints may pre-empt much of the dissatisfaction students experience when encountering them.

Students' skills at collaborative practice

All of the students in the study had experience of working in offline collaborations within their institutions. However, many of the students had not acquired the skills required for effective offline collaboration, such as working to deadlines and making timely responses to emails. The consequences of either displaying collaborative skills within the teams or not led to the development or breakdown of trust, echoing the observations of the project recounted in the previous chapter, and it appeared from the comments that acquiring a professional approach to university work is still not integral to the approach of all students. It also appears that far more collaborations in this study encountered problems with all students in the team showing commitment than in the previous study (which involved only two sites). This is accounted for simply by there being more students involved in the teams, and so the likelihood of teams including non-performers was raised. As noted in the previous chapter, once trust had been lost, it was rarely if ever rescued, as this is very difficult to do in an online environment and may have been exacerbated by the non-performing part of the team being more contrasted with the effective bilateral co-operation when this was successful. The general conclusion from this is that a three-way collaboration is very much more high risk than a two-way one.

The technology platforms

As noted in the previous chapter, technology adds transactional distance to the communication between participants. In our model of promoting effective online collaboration, a properly functioning technology platform needs to be introduced and ensured to be working well for online collaboration skills to be sufficiently underpinned. In this second iteration, a greater focus was made in investigating how students used the technologies available to them, hence its discussion here. Two factors bear on the extent to which the technology platform supports an effective learning experience for the students.

One of these is the intrinsic nature of the technology; both hardware and software presented issues in enabling collaboration effectively. The other is in the students' abilities in enacting appropriate behaviours to maximise the effectiveness of the technologies, that is, their abilities to adapt to problems with the technology, or make proper selections of the available technologies.

Technical issues with hardware

Four of the five groups said that they had issues with the hardware and software employed in their teamworking due to the specification of the devices used and the bandwidth of the networks. For example, the processing power of the computers used in some cases could not handle Word documents of the size the teams created. Having a dedicated machine was seen to be the solution for this.

> We still have the same issues with Loughborough students; they still have really bad connection, Whenever we have a meeting, within 10 or 15 minutes either we lose the sound or the picture. . . . The computer that our Loughborough students are using they seem to it was breaking down, because it crashed half way through one meeting and it crashed every time they shared the screen it happens.
>
> (CE1)

> technical issues in communication, sound, visual and connection. One day we could hear them, one day we could see them, one day we had to call it off because the connection was rubbish. Resolution: guarantee optimized platform, choose one computer and stick to it.
>
> (L1)

There should be standard stuff for hardware. If you're doing remote working then people need to have good microphones and good webcameras because otherwise and better internet connections because some of them have been really bad.

(CE1)

The final point is a good one; the project threw up the issue of the hardware support and bandwidth available is not sufficiently reliable at all places in universities to ensure communication can happen anywhere. Having dedicated machines is a solution in the short term, but this does seem to indicate that at many universities the available resources may not be sufficient to support online collaborative activities.

Issues due to incompatibility of software

Problems also occurred with the design software being used. Not all members of the teams used the same software, and there were compatibility and training issues in sharing documents between the different programs, partly because (as noted by Ryerson students) the Canadian university was using Apple Macintoshes and the UK universities were using PCs.

There are tools on a PC that are able to be used but they're not on our Macs. The drawing tools are on our macs are not on PCs so some people can use them and some people can't.

Compatibility across various products is a bit of a pain. The Ryerson students were using 2013 and I was using 2014 which was a bit of a pain. I had to downgrade. It was just a nuisance really. That's cross-working across different (institutions).

(CE1)

The comments of Loughborough students (following) very closely compare to those of Coventry and Ryerson students although they also note the lack of compatibility where they had expected different packages to be interoperable.

They should have been able to take the standard AutoCAD and open up in Revit. But if you had told me that in the beginning I would have done everything on Revit and learned how to use with a bit of help from the university so that you can then use these files rather than the day before you say it's not compatible. It was . . . things were not . . . so it looked uniform . . . so you spend the next 4 hours to make it look the same when you didn't know whether you were using the same program.

(L1)

However, there was another (though less common) response to the incompatibility of software that this was a learning opportunity, providing an occasion to learn from the other students about the different packages that were available.

I also thought that from a software viewpoint the multi-disciplinary team allocation allowed all members to show an interest in different software packages. Personally I was very interested in the CAD packages that the architectural students were using from Ryerson. The capabilities and end products that the applications produced were excellent; this experience was new to me as I haven't used similar CAD packages during my development.

(Personal reflection – Loughborough student)

Ability of students to select appropriate software

Several things emerge from collecting the students' feedback on the choice of technologies. The first is the high degree of digital literacy evident from the choices most of the students made, selecting appropriate technologies for separate forms of communication, for example, social networking sites for immediate responses (usually Facebook), email for less time-dependent responses, GoTo-Meeting for synchronous meetings (this was the platform they were instructed to use) and Dropbox for sharing documents.

The only mis-step was that one of the groups used more than one platform for sharing documents to begin with before switching to Dropbox alone.

We used Facebook (to plan meetings) and it's easy to upload photos and documents and you know who's seen it. (However), there's no filing system so if you upload too many documents you lose track of them.

(Personal reflection – Loughborough student)

The difficulties with curating the large number of documents produced by the task continued into their use of Dropbox. However, the students adapted to this with the folder arrangements that Dropbox allows.

For the largest submission (Task 2) there where hundreds of individual files that all needed to be collaboratively worked on and updated by multiple parties. This meant setting out a filing system/working plan early on (Task 1) to ensure that there would be no confusion with older drafts of files or naming etc. An online file sharing website has been used throughout all the tasks so every group member had access to the latest content.

(Personal reflection – Loughborough student)

Students also switched from platform to platform when communication if one proved ineffective, for example, trying Facebook when they got no responses from emails.

File sharing and communication was effective, and various media were used. Facebook was used in order to reschedule meetings and discuss any small problems or issues with work, Dropbox was used in order to collect useful files and upload work, and GoToMeeting was used for all of our meetings.

(Personal reflection – Coventry student)

Facebook was seen by some students to have a role in developing a social inter-action and helped to create the sense of togetherness that supports collaboration.

> Facebook had provided some sort of camaraderie.
>
> (Personal reflection – Coventry student)

Skills issues with effectively implementing technology

The students also reported some problems with learning to use the desktop shar-ing aspect of GoToMeeting, suggesting that they should be given some training or a short guide to its use.

> For the GoToMeeting it would be helpful to have like a quick introduction that tells you how to use it. Like, I'm quite good with Skype and I know how to use it from work. But not quite understanding the sharing your screen and stuff. It would have made it easier. Just an introduction of how to do it. have something simple like 2 minutes.

They also reported the problem with echo, indicating that the use of headphones to prevent this is not a normal part of their use of videoconferencing and had not occurred to them.

> I think the problem is having speakers on full blast right next to the micro-phone because you have an echo.
>
> (L1)

Similarly, the response of students to poor bandwidth in locations was to endure it rather than attempt to resolve the problems.

> here in the library for example we tried to have a meeting in one room in the library. We couldn't make any sense of the conversation. I kind of expected the internet connection to be bad.
>
> (CE1)

Observations on the role of the technology platforms

Having an appropriate set of technology platforms in place is obviously crucial for the effective deployment of online collaboration. Effectively engaging with the technology itself is also a pre-requisite for acquiring the skills required for the more specialised skill of collaborating online.

Activities such as online communication, particularly synchronous online com-munication, extend the requirements of institutions' information technology infra-structures beyond those of demands usually placed on it. Higher bandwidths are needed to ensure uninterrupted meetings, and faster processing power is required of laptops on which several programs are running simultaneously, such as GoTo-Meeting, Dropbox and AutoCAD. Students, however, seemed inexperienced at dealing with poor performance in the technologies they were using. When laptops were not of a high enough specification, there was no attempt to replace these with higher-specification ones. When meetings were held in locations where Wi-Fi was poor, there was no intent to relocate for future meetings. Furthermore, the echo, which can be heard on many recordings of meetings, was simply put up with. Echo occurs when someone speaks at location A and their voice is emitted from a speaker at location B and then picked up by microphones at that location and then sent back to location A. The speaker therefore hears their own voice back, with a short lag. Continuing under these conditions is extremely difficult and in professional settings is never tolerated. Some videoconferencing software uses echo cancellation to remove this, but it can also be resolved by participants using headsets or by turning off their microphones when they are not speaking. Like the poor performance, however, the students passively accepted the difficult circumstances.

With the software usage, however, students seemed far more willing to adapt and optimise their interactions. The switching between different means to curate online documents and select the most appropriate communication platforms for their needs showed a high level of digital literacy. The non-standardisation of software is something that ideally would be addressed at the institutional support level, with all contributing institutions identifying a common set of packages and ensuring that all students are trained in their use.

The disparity between students' high self-efficacy with respect to software and low self-efficacy with respect to hardware was unexpected and seems contradictory. A possible explanation is that, in their experience, hardware meets their needs without adaptation and is homogenous, whereas software frequently is incompatible and requires updating and adapting and reinstalling. With online collaboration, adapting the environment and technical setup of equipment to optimise the experience improves the quality of the collaboration. Making students aware that these can and should be optimised needs to be made clear, as this is apparently not self-evident.

Skills for online collaboration

Although the students had all experienced collaborative activities in earlier parts of their education, these had all been in face-to-face situations. Moreover, the experiences were with close colleagues, often friends, as collaborative groups were chosen along social lines. This meant that many of the standard aspects of collaborative working, such as scheduling and structuring meetings (assigning a chair, taking minutes), task allocation and decision making, had not been developed, as the majority of the tasks in the collaboration could be addressed in an *ad hoc* manner due to their frequent social contact. The distanced nature of the collaboration therefore not only meant adapting to an online environment, it meant acquiring many of the skills normally associated with face-to-face collaborations but in which the collaborators are not in daily contact with each other.

Skills associated with offline working that had not yet needed to be developed by students

Some groups did not use meeting management techniques effectively, for example, reflecting within meetings to clarify understanding and noting down action points.

> Some students don't understand what they have to do exactly. Even if the information is somewhere from out of the meetings, they didn't write it down or they forgot it.
>
> (CA1)

Management of deadlines and matching to competing demands at the different institutions was made difficult by the different module schedules described earlier but also was exacerbated by poor project management, with planning intermediate deadlines and workflow timing.

> Now according to the schedule that they (Loughborough)'ve got for us we've only got two days to do the calculations. How can we do the calculations in two days?
>
> (CE1)

Planning workflow also revealed issues in lack of management skills, in that the sequence in which information needed to be set out was not fully thought through.

> We've got to put together a materials list when we don't know how we're going to construct the building because no-one's given us feedback on our construction.
>
> (CE1)

Others displayed poor management of the technology, for example not checking their emails regularly.

Issues with task allocation

A particular issue that arose for the students was that of assigning tasks. Although having worked collaboratively before, for the students this was the first time they had worked on multidisciplinary teams and with other institutions. The disciplines correlated with the institutions (Ryerson $\equiv$ architecture, Coventry $\equiv$ structural engineering, Loughborough $\equiv$ engineering project management), so some of the issues may have arisen due to the dual divisions this created. The authenticity of any undergraduate assignment is also inevitably limited by the fact that it has marks which have to be allocated to all (potentially) equally, whereas in the actual work situation the assignment is intended to model, separate partners would be contracted to fulfil different amounts of the project. Students also see the assignment as an opportunity to develop skills and saw themselves as all equally project owners.

So for example, the difficulty faced by the Loughborough students was that their specific discipline they were learning was to project manage, and to have done this would have given them a leader role in the team, which was seen to be against the preference of the other members, which was to share in this role.

> Our discipline is management rather than design. I know we should have the insight, but they are architects and they've been doing it for 5–6 years. And designers design, and contractors construct, and we feel that we were overseeing it and we were also organizing the group and that is effectively what we are going to do in the future
>
> (L1)

When it came to design, this was not their remit, and yet there were more marks allocated to the design aspects, which were not seen as their remit, either by the Loughborough students themselves or by other members of the team.

> The mindset that we should be involved in the design but we wouldn't go into the architects' office and tell them what to do. So you work together collaboratively but to a certain degree and we wouldn't be responsible for it.
>
> (L1)

Whereas the design aspects were seen as clearly the remit of the architects, many felt they should have some input into the project management, particularly as this was seen to be skills they had all developed and because the project management role was seen as a more senior one.

> We tried being a bit more collaborative in the beginning, we split up the roles based on personal experience rather than what the intended role of each university is, but we were more than over-ruled. All of us had done a very similar project last year so we all had experience of all the different bits and pieces that were involved in different phases but our experience was just washed away. Loughborough decided that they had more experience in their degree of doing more of that so they assumed they would be doing more of that automatically and we wouldn't have a hand in it. Other than just spellchecking. We're the maths monkeys and they're the management.
>
> (CE1)

The lack of high levels of collaboration did not always translate necessarily to lack of consideration; it merely meant that the groups had divided up the workload initially and thereafter paid little interest in the others' work.

They did consider when they did the design to make sure it's not too difficult in that we can do the calculations in one day or two days.

(CB1)

Skills that emerge due to international and interorganisational collaboration

Working across time zones continued to be an issue across the entire activity, particularly with students not being aware that Daylight Saving ended at different times in the different countries.

You have to remember to tell them that the clocks have changed.

(CE1)

Our meetings would be at 4. But because of the time zones there was a bit of issue in that so just improving communication.

(L1)

However, this was not always attributed to inability to adjust to time zone differences but also due to lack of punctuality or assigning sufficient importance to the meetings.

Nonattendance to meetings and say that we are all going to meet up at 4 pm on Thursday and someone would not show up again, again and again.

(L1)

Same with the punctuality problems.

(L1)

Again, this suggests that although students were used to working collaboratively, they were used to working with close friends in a face-to-face environment, in which meetings occurred easily on an *ad hoc* basis. Arranging and attending formalised meetings was not therefore a skill they had developed, and this emerged in the move to online working, even though this is normally a skill associated with face-to-face collaboration too.

Skills required for synchronous working

As stated in the introduction, the "togetherness" of collaborating online is problematic, and as anticipated, students varied in their experiences of whether this occurred or not.

I'll be honest I didn't enjoy meeting online, it was a lot harder to communicate and I don't enjoy seeing my face when talking to someone. I will turn off the webcam of myself in future.

(CC1)

Others reported that their experience was very effective.

> I personally felt that the primary communication tool worked efficiently throughout the process and the functions it could perform were practicable for the tasks we were completing. Most importantly I would recommend this communication tool to industry professionals completing modern design projects based on my experiences.
>
> (Personal reflection – Loughborough student)

Through watching recordings of the students' videoconferences the development of this shared situational awareness can be observed. This developed in concert with a growing competency with running meetings (using agenda, assigning a chair, taking minutes by an assigned person, clarifying action points at the end), a growing dexterity with using the software (moving between programs, interacting with programs) and a developing maturity of the project (ideas moved beyond the initial "messy talk" phase, remaining tasks became clearer, whom to assign tasks became more obvious). When asked if there was a causation between these factors, that is did one become easier because another had become easier, students responded that no, these were correlated, in that all developed separately over time but did not directly help each other. However, from observing the recording of the interviews, all seemed to contribute to flow, which in itself contributes to the shared situational awareness.

Specific techniques that added to this awareness:

- Getting the technology right. In early stages, students used laptops that were too low spec to run GoToMeeting and CAD software simultaneously; students did not seek out the areas with better broadband connections; students did not use headsets or left microphones on when not speaking, leading to disruptive echoes as the output from mics were picked up by speakers. Some students adapted their surroundings to improve the experience, although as noted, many did not exercise this degree of control.
- Acquiring fluency. The more experienced the students became, the more able they were to switch between applications, zoom in on the parts they needed to zoom in on to illustrate a point and overall identify the best way to communicate visually autonomically, with the technology being transparent to their communication.
- Improving netiquette. Occasionally students needed to be preoccupied with some other task, such as searching on the Internet, looking for information on plans, opening files. In early stages students would simply withdraw from conversations, not answering questions until they had information and leaving other participants confused as to whether they were hearing them or were still present. As they became more familiar with the interactions, the students would explain what they were about to do before undertaking it so others would be aware of what was going on.
- Representing their thoughts visually. Students would describe parts of the building under discussion or discuss work they had undertaken, leaving the

screen with a static image and always leading to confusion with the other delegates. The students would then open a plan and use this to clarify what they were talking about. In later meetings, students immediately showed the plans and zoomed in on the part they were talking about. The others' attention could be drawn to the part of the diagram they were talking about using the cursor or even circling the specific part. More experienced students always backed up their communications with documents, for example demonstrating work they had undertaken, such as scans of written calculations or Excel spreadsheets.

- Increasing their presence. Students were very reluctant to create a presence on screen. At best all that existed of the student was the cursor moving on the screen, with the occasional flash of the webcamera image. Students communicated with each other using Facebook, which also would enable them to engage socially, but there was little or no socialising in meetings. Participants tended to refer to the members of the team at a distance by the institution at which they worked rather than by name. Having webcameras on view constantly (perhaps on a separate monitor, e.g. running a social channel in parallel) would enhance this aspect of collaboration.

- Expressing themselves on screen. Scribbling on diagrams and providing a quick diagram in SketchUp to explain a shape were all ways to express ideas and back up speech, but in a less formal manner. Having a whiteboard to hand to add an extra component is useful.

- Modifying each others' work. Where the students completely failed to engage with co-creation was by taking each others' documents and altering them. Although genuine and full co-creation did take place by commenting and requesting advice, the others' input was always added offline and outside of the synchronous sessions by the owners of the documents. This could be because of altering someone else's work being seen as intrusive or (because of the nature of the assessment) each wanting to keep their own contributions separate to some extent. Participants need to give each other permission to alter each other's work before they will feel comfortable with doing so.

In summary, in initial stages students gave little thought to the perception by their distanced peers about how their communication appeared on screen; there was little joint working or idea of the on-screen space being the shared space; activity too frequently broke off into the separate offline spaces. As the sense of the meetings being a shared situated experience developed, however, the majority of students' attention was on the shared collaborative space, indicating not only that a better collaborative experience had been formed but also potentially itself establishing a stronger shared collaborative experience. Several groups had recorded meetings which showed this transitory stage in which the mental location of the students switched from being situated in their separate locations to being within the shared virtual space.

Further development occurred over time, for example when comparing the recordings of videos in semester 2 to those in semester 1:

1   Meetings were shorter. The recordings made of semester 1 meetings are an hour long on average; semester 2 meetings are only 28 minutes long.
2   Meetings were more dynamic and focused. Switches between applications take place every 2 mins in one meeting, every 5 mins in another. This is a contrast with semester 1, when the same image sometimes stayed on the screen for up to 25 mins.
3   Information was shared effectively asynchronously, which means that less needs to be shared during the meetings. In fact one meeting ends with the statement that "we don't need to have another meeting because we are sharing emails so fast".
4   Participants were invited to ask questions at the end of presentations by name to ensure everyone got to speak.
5   The videos show higher social commitment to each other in general. In the majority of the videos, students thank each other for the promptness in replying to emails and suggest keeping in touch beyond the end of the project (possibly working on further projects), but one group selected to submit the recording in which their teammates did not show up.

In interviews, these observations were confirmed by the personal reflections of the students. First, that the second phase was conducted more effectively due to the established working relationships between Coventry and Loughborough.

> The second phase of the project was a lot more successful than the previous phase, due in part to having already had a working relationship with the Ryerson students. The task was more challenging as we were less familiar with the areas covered by some of the tasks, but the communication was a lot more effective due to the previous phase.
>
> (CE2)

Meetings were also conducted more smoothly because of the familiarity with the tasks and because the building design was more developed.

> We were able to communicate effectively with the Ryerson students, as we were all familiar with the design of the building and the construction process.
>
> (CE2)

The students were also more experienced at working across time zones and with collaborating online.

> Having worked together in the past we were able to quite easily overcome the challenges associated with not being able to meet in person, as well as

those met by living in different time zones. A routine for meetings existed from the previous phase and we were able to easily carry it over and incorporate the new members of the group into the process. From the previous phase we had learned what information was necessary to share and what was not, and consequently we were able to quickly share relevant information required for the furtherance of the project. This helped greatly speed up the meeting process, while still allowing us to communicate the same amount of information.

(CE2)

The students were also more effective at sharing information asynchronously, requiring synchronous communication just to iron out problems and to discuss next steps.

Overall, the communication throughout this phase of the project was of a good standard. Problems arising from the work were overcome via meetings on GoToMeeting, or, if they were smaller issues, were discussed via email or Facebook. The team were able to share ideas and solutions, and the existing working relationship and knowledge of the design was extremely useful for aiding the project.

(CE2)

Communication with the existing Ryerson team continued pretty much as before in the previous phase, using GoToMeeting and Facebook messaging for any contact outside agreed meeting times.

(CD2)

Also, for this phase we didn't use formal Meeting Minutes, we just posted all the information that needed to be finished in advance on our Facebook group since we already were familiar with this option.

(CB2)

The process was also simplified because of established schedules.

Learning from the first phase, the team agreed early on to have a regular set time to meet each week as we now had a good idea of when people were free. Continuity in team make up helped in this regard enormously.

(CD2)

And also familiarity with each other's abilities, which enabled easier division of responsibility.

Since we knew our roles from the beginning we didn't need to go through this all over again, we just divided the tasks according to each person's role.

(CB2)

Also the technical skills that had been acquired enabled more effective communication.

> Familiarity with the software available was a boon – technical hiccups were removed in entirety in the meetings with just the RU and CU students. We were only really aware of how smooth our use of the meeting software had become during the first meeting we had with the LU students – they had neither the appropriate hardware nor location to be a productive and helpful partner during the two meetings we had with them.
>
> (CD2)

As had resolving the technical difficulties in collaboration.

> RU and CU students struggled in the first stage with AutoDesk program compatibility as different editions do not work with each other. This sort of thing had been cured by the second stage.
>
> (CD2)

Greater familiarity with online behaviour also helped with effective communication, for example, learning communication techniques that compensated for the lack of body language in videoconferencing.

> I felt like in the first stage there were extended periods of silence in the meetings as team members weren't sure who would or could speak next – I think that might have been as a result of the reduced ability to read body language as one might do in person. This we had largely eradicated in the second stage by getting into the good practice of using an agreed meeting agenda, sent around the team members (usually) the day before the planned meeting. This produced more purposeful meetings and a more professional feel to them. There was also more of the life-like banter one might find in a meeting, due to better interpersonal relationships established over time.
>
> (CD2)

This ease of working with the others was not only manifested in the synchronous communication but in the asynchronous too.

> I felt that personally I had improved in communicating with other team members. This was on account of better familiarity with software but more importantly better relationships with team members. This was lacking in the first stage I felt, as it was somewhat awkward when requesting progress or information from other team members. The second phase saw that sentiment removed as I could flick a message to RU and not feel that it might be perceived rude and vice versa, RU students could send me a message out of meetings that was brief but to the point and it wouldn't feel brusque as we would catch up in the next meeting.
>
> (CD2)

Alterations due to addition of new LU students

Between semester 1 and semester 2, the membership of the teams stayed the same with regard to the Coventry and Ryerson members, but new students came in from Loughborough. Not only were these new to the activity, they were postgraduate students rather than undergraduates and were given the role of consultants to the existing teams. This was intended to maintain the activity as a three-site collaboration but presented a range of issues for some of the teams. In some recordings of meetings it can be seen that the Loughborough students are sharing materials and explaining the contribution they can make to the team, in others they are not present. The reduced size of the group helped for some.

> As a result of this effective reduced group size, we were able to proceed in meetings at good pace, not wasting as much time in meetings.
>
> (CD2)

> Only students from Coventry University and Ryerson University had to finish the rest of the project. Due to this fact, to communicate with each other was easier.
>
> (CB2)

Although others found this added to the difficulties.

> Now with less people to complete work and fewer minds to solve problems we were less of a team.
>
> (CC2)

The change in structure of the teams introduced some issues for the teams already set up. The first of these was that the students were unfamiliar with the project up to that point.

> Unfortunately the new members of the group were unfamiliar with the construction, and time that could have been used gathering information for the continuation of the tasks was instead used on explaining the design and background to the task. While the idea of a consultant was useful, it would have been more effective had we continued with the old team members who were familiar with the background to the project.
>
> (CE2)

> The team took time to readjust, people's motivation seemed to be depleted and as a result communication was less effective. We even missed a number of meetings due to time differences, which only happened once in semester 1.
>
> (CC2)

This was particularly difficult for some due to the role of the new people being that of consultants. As stated earlier, in the videos of the meetings it is evident that some Loughborough students demonstrated the sort of service they could undertake for the others.

> Though the addition of new parties to the group posed a challenge to us all, I feel as though we were able to take it in stride and adapt to the change so that time was not wasted and we could proceed with the task at hand. Instead of focusing on the negative aspect of new group members being added (and the additional learning curve), we took the opportunity to ask the new group members about some sustainable initiatives that they had in mind for the project, and everyone cooperated so that both the graduate and undergraduate students could gather enough information to complete their respective tasks.
>
> (R2)

Some groups, however, found this an adjustment that was difficult to make.

> Initially, being a later entrant as sustainable consultants into the project design phase, it was difficult to get adapted to the service model. Members in the consultant team hesitated to meet online and showed no interest to participate in discussion, as their involvement was neither monitored nor compulsory
>
> (R2)

The teams also missed the experience of the previous cohort of Loughborough students.

> LU students experience in management and leadership was valuable asset and because we had lost this, the team was less able to evolve – it was as if we were in a meeting for the first time.
>
> (CC2)

The difference in experience also showed itself in the unfamiliarity with the technology.

> during the first meeting between the Consultants and the designers, LU team had some technical problems which needed some time to be solved due to the absence of an expert, hence leading to a short delay.
>
> (R2)

And the ability to manage time zone differences.

> At first, Loughborough team made a mistake about the time difference, hence they missed the meeting for an hour which is really a pity because it wasted Canada students an hour.
>
> (R2)

Some teams however adapted quickly to the change stating that "We instantly developed a good rapport with the members from the two teams".

(L2)

Comparison with face-to-face teamworking

The aspect of the project that the students felt was more effective than their regular project working was the opportunity to work with students at other institutions. This was felt to be a more realistic simulation of the working environment because this imposed the need to present themselves in an outward-facing professional identity to external people rather than to friends from within their own institution.

> Surprisingly (previous face-to-face teamwork) did not work as well as this. You have to do more work in your own office back at home but it feels like there is so much more information flowing through despite the fact that we don't see each other physically. In the previous years we had teamwork but we were seeing each other more often during the week but it wasn't as efficient in the project itself in delivering the goods. . . . I think it's the extra motivation. It's something different, it's new and you don't want to let down the university if anything else. As in yeah ok if something is particularly expected from you, like they expect us to do the calculations and do the structural aspects of the project we want to make sure that we do that to the best of our abilities and at the same time contribute as much as possible in the other areas just to show involvement and just to show commitment to the project itself. And it just scores you extra points, and it doesn't matter for your grade but it matters for the way the teammates perceive you and it's important because it's not someone we're going to see day in day out at the university here, it's someone from somewhere and we assume they are just as well committed to the project as we are. It's just mutual respect, I suppose. It's more professional; we're not yet in the career but it's definitely closer to that than just being with pals at the university.

(CA1)

The disadvantage is the lesser efficiency in working with people that are unknown, compared to people with whom they have already built up a working relationship.

> Those that didn't do it wanted to stay with their mates. They've created their own little group and that's what they want to do. They've stayed in that group for the past three years now. They work more efficiently this way by staying in their groups.

(CA1)

The added effort required by forming new groups for the virtual teamworking was seen as adding more authenticity to the exercise, as do cultural differences described earlier.

I know when I struggle when I get a job so I might as well get used to it. I'm expecting it to help.

(CA1)

At least when you get out into the real life you won't be shocked by what is happening. By what's happening now I'm like "I'm not used to this – oh I have to work with so many people. W so now you have this experience to actually work with someone who (has different practices).

(CA1)

Finally two of the groups (and in a quick poll, this was the opinion held by about half the students in the class) stated that "SCREENS ARE NOT ENGAGING" and "I still believe face to face meetings are key to success", that is, that conducting teamwork entirely virtually is not effective in itself and that effort should be made to enable face-to-face activity to take place.

Conclusions

The experiences of online learners throughout this project show enormous growth in abilities and in confidence throughout the two semesters. Despite lacking some of the basic elements of project management practices, these were soon developed during the activities and in most cases made the collaborations far more effective. Students were also lacking in many of the skills required to conduct synchronous communication effectively, indicating a lack of experience in these systems. The awareness of project needs and task requirements grew in parallel with fluency in operating the software in the synchronous meetings and a focus on providing efficient exchange of information in the asynchronous modes.

The growth of ability of the members of the teams in the first semester is particularly apparent when compared with the difficulties faced by the new members of the team in the second semester. In part this was due to lack of clarity about their role, but it also indicates the degree to which the sense of working together in a shared online space had occurred as the newcomers struggled to share that same sense of space.

Further conclusions, as well as drawing together observations from this, the previous chapter and ensuing chapters, are discussed in the final chapter of this book.

4 Designed to be employed? Measuring the impact of a multidisciplinary collaborative design project on learner perceptions of employability attributes

*Robby Soetanto, Mark Childs, Paul S.H. Poh,
Jacqueline Glass, Stephen Austin,
Zulfikar A. Adamu and Chinwe Isiadinso*

Introduction

A collaborative building design project undertaken within an internationally distributed team involves a dynamic process characterised by generation and sharing of information and synthesis of knowledge between participants. Learning within this dynamic environment is challenging but can bring a number of notable benefits for the participants. Inherent within successful collaborative learning is the required ability to co-produce design 'content' with others from different disciplines and to manage the 'relationship' among all participants involved in the design team (Leinonen, Järvelä and Häkkinen, 2005). The 'content' constitutes individuals' inputs, which originate from disciplinary knowledge, skills and expertise, whereas managing the relationship requires a set of 'soft' people management skills. During the design process, the participants are presented with a problem (in a building project, it is usually a client brief), which has multiple potential solutions. To arrive at an optimum solution, the participants should explore the rationale of each alternative, and present and negotiate alternatives with the other participants. This process encourages deep learning of the subject discipline and helps to develop people management skills, such as teamwork, communication and other performance-enhancing behaviours which have been linked to 'proactive personality' (Tymon, 2013). In the present and future labour markets, graduates are expected to be able to work across disciplinary and geographical boundaries (Becerik-Gerber, Ku and Jazizadeh, 2012; BIM2050 group, 2014), and these people management skills have been identified as the skills for developing sustainable built environment (BE) (Egan, 2004). In a report commissioned by the UNESCO, Beanland and Hadgraft (2014) further stressed the importance of the development of appropriate interpersonal attributes and capabilities as an integral part of engineering education worldwide.

The evaluation of the skills developed from the learning activity is critical to demonstrate the success of the learning endeavour and is reflected in the achievement of learning outcomes. There exist examples which explore and evaluate the impacts and benefits of collaborative design projects in the built environment (e.g. Becerik-Gerber, Ku and Jazizadeh, 2012), but none evaluate pedagogical and personal development skills from the learner perspective and then compare the developed skills before and after a learning intervention. This evaluation is presented in this chapter with a view to enhance understanding of the effectiveness of this learning approach and to inform key requirements of its successful implementation.

This chapter considers several issues in the evaluation of collaborative design project, including measures which allow consistency of evaluation of pre- and post-implementation across several disciplines involved in the collaboration, learning outcomes of each discipline and the influence of the participants' beliefs about the benefits of a collaborative design project.

Thus, key research questions are suggested as follows.

1   What are the impacts of a collaborative design project on learning outcomes, defined by attributes including understanding, ability, skills and qualities, as perceived by the participants, pre- and post-implementation?
2   Is there any difference in perceived impact between participants from different disciplines and those with different beliefs about the efficacy of the learning approach?
3   What attributes can facilitate successful collaborative design?

In the BIM-Hub project, the participating students were asked to rate their perceived ability against a range of understanding, ability, skills and qualities, defined as attributes at the start and on completion of the project. Data were obtained from 139 completed pre- and post-implementation questionnaires. The following sections review key literature related to building design projects and evaluation of key skills requirements before the detailed approach and research methods are explained. The findings are then presented and discussed. Conclusions are drawn to address the above research questions.

Evaluation of building design project and key skills requirements

The implementation of collaborative learning is believed to bring additional benefits of promoting 'soft' people management, employability skills, motivating and improving student engagement. Several studies, such as that by Thloaele, Suhre and Hofman (2015), provided evidence that technology-enhanced, cooperative group-project learning enhances academic performance. However, an evaluation of students' perceived skills resulting from working in a virtual collaborative design in BE disciplines is scant. This is rather surprising given that the idea of a collaborative design project is not new. Most published papers discussed the evaluation from the tutor perspective, highlighting the development of interpersonal skills (e.g. Becerik-Gerber, Ku and Jazizadeh, 2012). Selected published findings from the evaluation of collaborative design work are discussed in the following paragraphs.

O'Brien, Soibelman and Elvin (2003) examined student self-assessment of collaboration in a virtual, multidisciplinary design project between two universities. The results suggest that most students felt that the groups performed efficiently and produced high-quality designs. Some 20% of the students responded neutrally or negatively about the experience due to the difficulties encountered during the collaborative process. Individual critiques revealed that the difficulties stemmed from cultural differences between group members rather than from technologies or distance between team members. There was no evaluation of skills gained from the collaborative design exercise.

Tucker and Rollo (2006) reported on their experience of running the collaborative design module over three academic years. They found that the performance of participating students was higher than that of non-participating students. Lessons learnt from the previous years provided considerations for changing student grouping strategies, which had brought performance improvement for the whole cohort. This is inconsistent with the assessment results reported by Soetanto *et al.* (2014), who suggested that the collaborative design project has no impact on individual and group marks, but it does develop a proactive attitude among the participating students, highlighting 'proactive personality', a term coined by Tymon (2013).

Bhandari, Ong and Steward (2011) developed a multidisciplinary course on sustainable engineering design. They found that participating students felt that their ability to consider technical, environmental and social aspects of sustainability in development of engineering design solutions improved. From interviews with students participating in an integrated sustainable design experience, Wolcott *et al.* (2011) found that the students gained improved communication and teamwork skills. Stanford *et al.* (2013) statistically analysed student self-rated response of eight relevant sustainability concepts from pre- and post-implementation of a capstone sustainable design project. The analysis revealed significantly improved knowledge in six out of eight sustainability areas including economic analysis, sustainability rating systems, impact on society, life cycle assessments, sustainable transportation and sustainable materials. Further qualitative analysis of students' journals confirmed improved critical thinking skills. Korkmaz (2012) analysed qualitative responses of a pre– and post–case study assignment survey to evaluate the effect on student learning. The analysis showed that case-based collaborative-learning methods improved student learning about the delivery of sustainable building attributes.

Becerik-Gerber, Ku and Jazizadeh (2012) discuss the learning outcome of virtual collaborative design based on several primary sources, including a student satisfaction survey, a series of surveys completed by students at the end of each assignment, official course evaluations, formal/informal discussions with students, and instructors' evaluation of students' work. From the student survey alone, they found that modelling and virtual collaboration skills have adversely impacted their experience, and there was disparity between skills levels and abilities of students from each institution. Although they preferred to have in-person rather than virtual meetings, a virtual design project improves the learning of collaborative building information modelling (BIM) concepts.

Solnosky, Parfitt and Holland (2013, 2014) developed a multidisciplinary architectural engineering capstone design course to enhance student skills to work within integrated project delivery (IPD) and BIM environment. They mapped the 'short-term' educational objectives of the course against the educational outcomes, which are defined as noticeable traits that can be seen for up to five years after taking the course. The assessment of objectives and outcomes was undertaken by faculty members and industry practitioners. There is no evidence of student evaluation of the skills developed from the course.

Most evaluations were conducted by asking the experience of students, providing qualitative responses. The impact of the learning approaches on the participant attributes following implementation has not been fully assessed quantitatively. Despite a lack of consensus, several studies suggested perceived skills improvement, but it is unclear which skills and why they have been improved. For only a few comparisons between pre- and post-implementation (such as Stanford *et al.*, (2013), on student knowledge of sustainability), evaluation was based on the knowledge gained rather than transferable skills and attributes and was based on limited samples ($n = 19$ for pre- and $n = 15$ for post-implementation in Stanford *et al.* (2013). Most design projects (as discussed earlier) are not based on virtual collaboration, with exception of Becerik-Gerber, Ku and Jazizadeh (2012). Given the variability of implementations, lessons learnt are not directly applicable and comparable for a different context (Soetanto *et al.*, 2014). As there is a lack of longitudinal research, there is no control group to benchmark the skills improvement, and it is not possible to test the learners' skills in their professional life.

Learning outcomes of collaborative design

Assessment is an essential part of learning and determines how and what the students learn. To promote deep learning, the assessment tasks are designed to test whether the students have met these learning outcomes (Biggs, 1996; Ayres, 2015). A collaborative design project is built on a constructivist approach to learning, where the students build on their existing knowledge through the exploration of the subject, sharing and discussing their findings and ideas with the other group members (Vygotsky, 1978, c.f. Ayres, 2015). The students take responsibility for their own learning by actively undertaking learning activities, such as searching, exploring and sharing information with others, problem solving and discussing problems in their group. These activities developed higher-order thinking skills, such as analysis, synthesis and evaluation (Ayres, 2015). A collaborative learning process is seen as a means to achieve learning outcomes and should be considered as an important element in the assessment of the learning. Thus, the challenge is to develop an appropriate means to assess an individual's learning within a collaborative learning context.

The assessment of collaborative learning is different from that of individual learning because the learning occurs with a group environment. The assessment of individual learning with a collaborative learning context is found to be challenging due to the difficulty in completely separating the learning environment (i.e. the project, team dynamic) from individual student learning (Howard, 2014). The group working process and team dynamic can determine what individual members learn and achieve at the end. The final grade attainment of the group may not have any relationship with individual achievement. Thus, there may be cases where the group grade is high, but the individual only achieves a satisfactory level of learning. This may be because members try hard to maximise their group grade to the detriment of individual learning gain. In engineering education, quantitative assessment methods (such as exams) are a common way to assess summative learning of individual students, but these can lead to competitive and mark-chasing behaviours which do not really promote deep learning. Howard (2014) suggests that assessment of individual learning in a group environment should be undertaken via qualitative methods, such as portfolio and presentation.

Group work develops students' confidence to be an independent learner through taking responsibility for their own learning. It permits the students to develop their people management skills and an independent attitude and provides a sound platform to develop their knowledge, understanding and skills for subsequent years. Here, a successful outcome does not entirely rely on the technical prowess of individual members but depends on the ability of the team to work together and integrate the work into a coherent piece. These are the very skills that are in high demand by the industry, and hence a group assignment helps to develop employability skills (Ayres, 2015).

To determine what the students should achieve from collaborative learning activities, we need to consider the perspectives of multiple stakeholders (Tymon, 2013). Here, the needs of employers are, perhaps, among of the most important and have often been the aim of educational institutions. The term 'employability skills' has been used to describe the skills that make graduates employable at the end of their study. Tymon (2013) argued the emphasis of employability should be on the 'ability' rather than 'employment' or 'being employed'. The first term encompasses a larger scope of developing learning activities to equip graduates with enabling lifelong skills to contribute to society at large. In practical terms, employability skills are often meant to include a whole range of skills which make graduates more employable. Some argue the need to have other people management skills on top of the technical skills (BIM2050 group, 2014). Although it is rather contentious, the current discourse suggests that employability skills are the skills that give graduates a competitive advantage in the job market. Given the diversity of perspectives regarding the knowledge and skills of collaborative learning, a standard to assess the learning outcomes could help stakeholders to focus on important aspects of learning activities. For example, educators would be able to design learning methods, activities and assessments which address the outcomes. Standard learning outcomes could also raise learners' awareness of the expectations of the programme. In BE education, such standards are typically set by professional bodies and used in the accreditation of degree programmes. Due to multiple disciplines involved in a collaborative design, however, a review of international standards of relevant disciplines is provided in the following sections.

Engineering education standards

Standards in engineering education have received significant attention due to increasing perceived need to redefine the future education needs of practicing engineers, who are able to contribute to the society (Beanland and Hadgraft, 2014). Emerging issues and challenges, such as sustainability and climate change, have provided an impetus to broaden the scope of engineering education in order to equip the future engineers with relevant knowledge and skills to undertake their changing roles. Through accreditation, professional institutions have an important and influential role to specify the knowledge, skills and behaviours that graduates should be able to demonstrate on the completion of programmes (Beanland and Hadgraft, 2014). Hence, the learning outcomes of relevant professional institutions are described here.

An international standard for engineering education was developed by the International Engineering Alliance via the Washington Accord Agreement. The agreement is binding to the signatories' countries. It provides a professional competencies profile and graduate attributes associated with each of three categories of employment in engineering, i.e. Professional Engineers, Engineering Associates and Engineering Technicians. Graduate attributes are met when a student completes an accredited engineering degree programme. A professional competencies profile should be fulfilled five years later when the individual looks to apply for professional engineering registration (Beanland and Hadgraft, 2014). The Accord stipulates that two fundamental objectives of engineering education are to build a 'knowledge base' and 'attributes' of the graduates to continue learning and develop competencies required for independent practice. The Accord sets a minimum standard which graduates from any accredited engineering degree programme should achieve by the end of their study. Perhaps the significance of the Accord is to provide assurance that graduates from accredited programmes from signatory countries could operate together in a project team.

Construction management education standard – the Chartered Institute of Building (CIOB)

In the UK, the standard of construction management education is based on the CIOB education framework (CIOB, 2013). The framework provides guidance to teaching institutions to review the existing programme content, a reference document when designing a new programme and, importantly, for gaining accreditation for the programme. Although it is not meant to be a prescriptive document, the CIOB accreditation is based on an education framework and provides an approval that teaching institutions or programmes have met the highest standard of quality in construction management education. The education framework was developed based on UK Quality Assurance Agency benchmarks and National Occupational Standards. The framework is applicable to four educational levels from diploma to postgraduate programmes. Learning outcomes of the education framework are grouped under six main themes, namely sustainability, construction environment, construction management, construction technology, health, safety and welfare, and ethics and professionalism. They are further detailed in sub-themes, and mapped against educational levels.

Although the skills and attributes associated with collaborative design are not obvious from the main themes, there are some detailed items which underpin knowledge and behaviour to work in collaborative design, including 'team and integrated teams', 'building information modelling (BIM)', 'respect for other professions and their roles' and 'internationalisation'.

Architectural education standard – Canadian Architectural Certification Board (CACB) and Royal Institute of British Architects (RIBA)

Accreditation of architectural education in Canada is undertaken by the Canadian Architectural Certification Board (CACB, 2012). The board stipulates 31 student performance criteria which define the skills and knowledge for all graduates of an accredited degree programme. The criteria are organised according to four categories, namely 'critical thinking and communication', 'design and technical skills', 'comprehensive design' and 'leadership and practice'. Several relevant criteria for a collaborative design project include 'collaborative skills' and 'leadership and advocacy'. Both determine the skills to work collaboratively with other disciplines in a project and maximise outcomes through these skills.

The accreditation of architectural courses in the UK is undertaken by RIBA via its Education Department (RIBA, 2014). RIBA (2014) stipulates that students in accredited architectural courses should be able to demonstrate graduate attributes at parts 1 and 2 before Professional Practice Examination in Architecture (part 3). Similar to CACB's attributes, several attributes are relevant to a collaborative design project, summarised as: 'knowledge of professional inter-relationship of parties', 'ability to apply a range of communication methods and media', 'knowledge of context and industry', 'professional quality and problem-solving skills', 'professional judgement', and 'ability to take the initiative and make appropriate decisions', 'skills necessary to plan project-related tasks' and 'coordinate and engage in design team interaction'.

BIM learning outcomes

The UK's BIM Academic Forum (BAF, 2013) established learning outcomes from BIM learning. The learning outcomes are grouped in three main categories, namely 'knowledge and understanding' (including principles and concepts of BIM), 'practical skills' (ability to use BIM software, e.g. Revit) and 'transferable skills' (including collaborative working, communication in multidisciplinary team). The categories are cross-referenced against educational levels from first to final years of undergraduate and postgraduate programmes. This sets an expectation of what each level should achieve for each category.

At year one (level 4) of undergraduate study, the learning outcomes are to provide the context and background to the industry, emphasising students to appreciate the manner in which the industry works, the roles and disciplines involved, to introduce how information is prepared, shared and issued and technologies to support BIM and promote collaborative working. For year two (level 5), the learning outcomes are to develop the knowledge and understanding of the role of BIM as a business driver for collaborative working within an integrated supply chain. Year 3 (level 6) learning outcomes should focus on building competencies and knowledge around people, systems and processes of BIM. The students should develop an awareness and appreciation of the cultural and organisational impacts of BIM, an appreciation of new ways of integrated team work and collaborative working environments (BAF, 2013).

In summary, this review suggests different standard learning outcomes across different BE disciplines. It is worth noting that all recognise the essential attributes required for collaboration in project teams. Given that activities in collaborative learning are self-regulatory, the perceived learning gain may be influenced by the participants' belief that working on a collaborative design project will improve their attributes. The underpinning theory is explained in the following section.

Self-efficacy theory

Self-efficacy is a theory in educational psychology which focuses on establishing a relationship between learners' belief of what they can achieve and the outcomes of learning. Bandura (1977) first coined 'self-efficacy belief' as an important attribute of performance in learning. He defined perceived self-efficacy as personal judgments of one's capabilities to organise and execute courses of action to attain designated goals. Self-efficacy is a multidimensional concept in itself but also relates to other concepts such as outcome expectation, self-concept and perceived control. Nevertheless self-efficacy focuses principally on a specific task that the learner is going to undertake (Zimmerman, 2000). Self-efficacy belief was found to be a significant predictor of academic performance and was positively correlated with motivation, participation, persistence and the amount of effort. Self-efficacious students tend to be more proactive and take initiative in their group learning (i.e. self-regulated learning, defined by Bandura, 1977). Given that the activities involved in collaborative design project are self-regulatory, there is potential impact of self-efficacy belief on the evaluation of learning outcomes. This was considered in this research by asking participating students whether they believe that working on a collaborative design project will enhance their attributes. The purpose is to identify whether having self-efficacy belief will impact the evaluation pre- and post-implementation.

Methods

Attributes of knowledge, understanding, ability, skills and qualities

The engineering subject benchmark contains statements for knowledge, understanding, ability, skills and qualities that were considered necessary to enable graduates to work effectively in their professional careers (Maddocks, Dickens and Crawford, 2002). They were subsequently termed 'attributes'. Although they were the benchmark for engineering degree programmes, the attributes contained within are reasonably general to be adapted for evaluating achievement of students participating in a collaborative design project. The intention was to cover as many attributes as possible within the subject benchmark, because collaborative design can potentially contribute to the development of a range of graduate attributes. However, 'ability to use laboratory and workshop equipment to generate valuable data' was removed, as it was deemed to be not relevant. The number of attributes included in the questionnaire was considered to be reasonable for a short (10–15 minutes) completion but also sufficient to allow an evaluation of a wider range of learning gains afforded by a collaborative design project. The attributes representing skills, abilities and qualities are categorised under five headings:

- Knowledge and Understanding
- Intellectual Abilities
- Practical Skills
- General Transferable Skills
- Qualities

Before their inclusion in the questionnaire, the statements were adapted to more concrete terms to make them more relevant and allow participating students to effectively consider them within the collaborative design project. The attributes are listed in the second column of Table 4.1.

Questionnaire design and distribution

Participating students were asked to indicate their assessment of their ability for each attribute using a scale from 1 to 4, where 1 indicates 'poor', 2 'moderate', 3 'good' and 4 'very good'. They were also asked to indicate their perception of whether working on a collaborative design project would improve their achievement on each attribute. They indicated their opinion through a binary answer, 'yes' or 'no'. The responses allow an exploration of the impact of self-efficacy belief on the learning gain in a collaborative design project. For both questions on attributes and perceived benefits, a 'do not know' answer was provided. The questionnaire simply lists the statements without headings, as this presentation prevents the students giving the same rating to attributes in the same heading. Following completion of questionnaire design, human participant ethical approval was obtained.

Questionnaires were distributed and collected by hand in the class, first in the second week and second in the penultimate week of their involvement with the activity. The participating students were given about 10 to 15 minutes to complete the questionnaire.

Analysis of data

Data from the completed questionnaires were coded and input into SPSS version 22. Pre- and post-implementation scores were averaged and the discrepancies

Table 4.1 The results of statistical analysis of pre- and post-implementation scores and the influence of the perceived benefits and discipline groups

Code	Skills/attributes/qualities	Average score		Difference	Ranking of the most improvement	t-test (Sig, 2-t)	ANOVA (Sig.)	Two-way ANOVA	
		pre	post					Stages vs. belief	Stages vs. disciplines
A01	Ability to demonstrate sound specialist/discipline knowledge	2.81	3.02	0.21	13	0.087*	0.000**	0.611	0.497
A02	Ability to demonstrate understanding of external constraints	2.81	3.06	0.25	9	0.021**	0.000**	0.171	0.795
A03	Ability to demonstrate business and management techniques	2.72	2.67	−0.05	25	0.675	0.000**		0.604
A04	Ability to demonstrate understanding of professional and ethical responsibilities	2.88	3.05	0.17	17	0.174	0.000**		0.385
A05	Ability to demonstrate understanding of the impact of construction/engineering solutions on society	2.74	3.03	0.29	4	0.039**	0.000**	0.384	0.351
A06	Ability to demonstrate an awareness of relevant contemporary issues	2.51	2.71	0.20	14	0.104	0.000**		0.499
A07	Ability to solve engineering and design problems through creative and innovative thinking	2.73	3.02	0.29	4	0.038**	0.000**	0.642	0.330
A08	Ability to apply mathematical, scientific and technological tools	2.68	2.92	0.24	10	0.095*	0.000**	0.000**	0.137
A09	Ability to analyse and interpret data and, when necessary, design experiments to gain new data	2.47	2.69	0.22	11	0.104	0.000**		0.765

(Continued)

Table 4.1 (Continued)

Code	Skills/attributes/qualities	Average score		Difference	Ranking of the most improvement	t-test (Sig. 2-t)	ANOVA (Sig.)	Two-way ANOVA	
		pre	post					Stages vs. belief	Stages vs. disciplines
A10	Ability to maintain a sound theoretical approach in enabling the introduction of new technology	2.49	2.69	0.20	14	0.131	0.000**		0.466
A11	Ability to apply professional judgement, balancing issues of costs, benefits, safety and quality	2.85	2.90	0.05	23	0.673	0.000**		0.521
A12	Ability to assess and manage risks	2.92	3.05	0.13	20	0.204	0.001**		0.579
A13	Ability to use a wide range of tools, techniques and equipment (including software) appropriate to their specific discipline	2.72	2.89	0.17	17	0.182	0.000**		0.227
A15	Ability to develop, promote and apply safe systems of work	2.78	2.87	0.09	22	0.499	0.001**		0.622
A16	Ability to communicate effectively, using both written and oral methods	3.03	3.25	0.22	11	0.058*	0.000**		0.002**
A17	Ability to use information technology effectively	2.89	3.24	0.35	1	0.006**	0.005**	0.031**	0.011**
A18	Ability to manage resources and time	2.91	3.18	0.27	7	0.019**	0.000**	0.244	0.508
A19	Ability to work in a multidisciplinary team	3.11	3.39	0.28	6	0.011**	0.000**	0.280	0.269
A20	Ability to undertake lifelong learning for continuing professional development	2.89	3.15	0.26	8	0.054*	0.000**	0.315	0.782
A21	Creative, particularly in the design process	2.66	2.84	0.18	16	0.230	0.000**		0.515

(Continued)

Table 4.1 (Continued)

Code	Skills/attributes/qualities	Average score		Difference	Ranking of the most improvement	t-test (Sig, 2-t)	ANOVA (Sig.)	Two-way ANOVA	
		pre	post					Stages vs. belief	Stages vs. disciplines
A22	Analytical in the formulation and solutions of problems	2.67	2.97	0.30	3	0.012**	0.000**	0.002**	0.287
A23	Innovative, in the solution of engineering problems	2.71	2.87	0.16	19	0.253	0.000**		0.286
A24	Self-motivated	3.27	3.31	0.04	24	0.730	0.000**	0.417	0.036**
A25	Independent of mind, with intellectual integrity, particularly in respect of ethical issues	2.79	3.11	0.32	2	0.018**	0.000**		0.476
A26	Enthusiastic, in the application of their knowledge, understanding and skills in pursuit of the practice of engineering	2.93	3.06	0.13	20	0.299	0.000**		0.925

Note: * significant at 10%, ** significant at 5%

between them calculated. The discrepancies indicate the amount of improvement in student ability before and after participating in the project. They were then ranked in descending order of magnitude, as presented in Table 4.1, column 6. To confirm the difference between pre- and post-implementation scores, the data were subjected to an independent samples *t*-test. The grouping (independent) variable was stages (pre- and post-implementation); the test (dependent) variable was the score of each attribute. The *t*-test yielded probability (*p*-value) of obtaining the results due to sampling error if there was no difference between pre- and post-implementation scores. The threshold for confirming the difference (level of significance) was set at 5%, deemed to be appropriate for quantitative data analysis in social science and applicable to statistical analysis presented in this chapter. The *p*-value for each attribute is presented in Table 4.1, column 7.

The scores were subjected to analysis of variance (ANOVA) to identify any differences of scores between discipline groups (architect, civil/structural engineer, construction manager, postgraduate project manager). The results are presented in Table 4.1, column 8. A two-way ANOVA was employed to identify interaction between stages and discipline groups, and stages and perceived benefit of the project on the scores. The interaction between stages and perceived benefit investigated whether the scores of perceived ability in the pre- and post-implementation were influenced by the perceived benefit of participating in the collaborative design project. It is hypothesised that those students who perceive that working on a collaborative design project is going to help them improve their performance will perceive higher performance in the post-implementation. The interaction between stages and discipline groups investigated whether the scores of perceived ability in the pre- and post-implementation were influenced by their membership in discipline groups. In other words, it investigated whether the perceived level of performance pre- and post-implementation is the same or otherwise across different discipline groups. The test of the interactions was conducted to only those attributes with statistically significant (in the *t*-tests and ANOVA tests) because they are considered worthy of further investigation. The results of two-way ANOVA are provided in Table 4.1, columns 9 and 10.

Principal component analysis (PCA) was used to explore the grouping of required attributes to support a successful collaborative design project. The use of PCA rests on the assumption that the attributes would cause or produce the components (rather than that the components cause attributes), hence the exploratory nature of the analysis. The Kaiser-Meyer-Olkin measure of sample adequacy of 0.912 and Bartlett's test of sphericity with $p < 0.0005$ indicate that there are correlations between attributes, and so PCA can be meaningfully applied. The number of components was determined based on the number of components that have an Eigen value of more than one. This was supported by the observation of the scree plot. Oblique rotation (Promax with kappa of 4) was adopted based on the assumption that groups of attributes are likely to be inter-related in a real context (Male, Bush and Chapman, 2011). Promax with kappa of 4 was applied with the intention that it can facilitate a clear grouping of attributes into their components. The finding is summarised in Table 4.2.

Table 4.2 Summarised results of principal component analysis

Code	Skills/attributes/qualities	Component loading	Variance explained (%)
Component 1: Teamwork, management and motivation			**42.552**
A16	Ability to communicate effectively, using both written and oral methods	0.757	
A17	Ability to use information technology effectively	0.683	
A18	Ability to manage resources and time	0.634	
A19	Ability to work in a multidisciplinary team	0.841	
A20	Ability to undertake lifelong learning for continuing professional development	0.637	
A24	Self-motivated	0.733	
Component 2: Technical knowledge for creativity and innovation			**6.348**
A01	Ability to demonstrate sound specialist/discipline knowledge	0.631	
A07	Ability to solve engineering and design problems through creative and innovative thinking	0.694	
A08	Ability to apply mathematical, scientific and technological tools	0.716	
A09	Ability to analyse and interpret data and, when necessary, design experiments to gain new data	0.688	
A10	Ability to maintain a sound theoretical approach in enabling the introduction of new technology	0.623	
A13	Ability to use a wide range of tools, techniques and equipment (including software) appropriate to their specific discipline	0.677	
A21	Creative, particularly in the design process	0.811	
A22	Analytical in the formulation and solutions of problems	0.761	
A23	Innovative, in the solution of engineering problems	0.668	
A25	Independent of mind, with intellectual integrity, particularly in respect of ethical issues	0.610	

(Continued)

Table 4.2 (Continued)

Code	Skills/attributes/qualities	Component loading	Variance explained (%)
Component 3: Understanding implications and enthusiasm			**5.061**
A03	Ability to demonstrate business and management techniques	0.720	
A11	Ability to apply professional judgement, balancing issues of costs, benefits, safety and quality	0.718	
A12	Ability to assess and manage risks	0.708	
A15	Ability to develop, promote and apply safe systems of work	0.755	
A26	Enthusiastic in the application of their knowledge, understanding and skills in pursuit of the practice of engineering	0.665	
Component 4: Knowledge and understanding of wider context			**4.535**
A02	Ability to demonstrate understanding of external constraints	0.849	
A04	Ability to demonstrate understanding of professional and ethical responsibilities	0.617	
A05	Ability to demonstrate understanding of the impact of construction/engineering solutions on society	0.683	
A06	Ability to demonstrate an awareness of relevant contemporary issues	0.682	
Total			**58.496**

Findings and discussion

Perceived achievement before and after the implementation of collaborative design and in discipline groups

Table 4.1 (columns 3–6) presents the average scores pre- and post-implementation, the differences between pre- and post-implementation average scores and the ranking of the most improved scores. The differences between pre- and post-implementation scores show that all but one of the attributes have been perceived to be improved after participation in the collaborative design project. This is a positive outcome, endorsing the benefit of learning through a collaborative design project. 'Ability to use information technology effectively', 'Independent of mind, with intellectual integrity, particularly in respect of ethical issues' and 'Analytical in the formulation and solutions of problems' were the three most improved attributes, followed by 'Ability to demonstrate understanding of the impact of construction/engineering solutions on society' and 'Ability to solve engineering and design problems through creative and innovative thinking' as joint fourth ranked. These improvements were confirmed by a *t*-test as presented in Table 4.1, column 7, which indicated that the difference between pre- and post-implementation average scores of all five attributes was statistically significant. The other significant attributes include 'Ability to work in a multidisciplinary team', 'Ability to manage resources and time' and 'Ability to demonstrate understanding of external constraints', which were ranked sixth, seventh and ninth, respectively. Despite being ranked eighth, the attribute 'Ability to undertake lifelong learning for continuing professional development' was not statistically significant at 5%, and therefore, this result only provides some evidence of improvement.

At the bottom of the ranking, 'Ability to demonstrate business and management techniques', 'Self-motivated' and 'Ability to apply professional judgement, balancing issues of costs, benefits, safety and quality' were ranked the weakest three attributes. The average score of one attribute, 'Ability to demonstrate business and management techniques', was found to be lower after the implementation. Perhaps the link between the collaborative design activities and business and management technique was not clearly established. It is worth noting that 'self-motivated' was the attribute with the highest average score for both pre- and post-implementation. This suggests that the motivation of participating students remains high throughout the implementation of the collaborative design project, confirming the benefit of collaborative learning for motivating and engaging students.

The results indicate that the students experienced significant improvements in their abilities to use ICT for collaborative design tasks; deploy creative, analytical and innovative thinking to solve engineering and design problems; understand the impact of construction/engineering solutions and any external constraints; manage their resources and time; and demonstrate independence of mind with intellectual integrity. Also, having participated in the project, the students became more acquainted with working in a multidisciplinary team, recognising the significance of leadership, and they better understood their role in a team.

The results of investigating the perceived achievement of discipline groups show that average scores were significantly different among groups (based on ANOVA tests, presented in Table 4.1, column 8). This indicates that each group perceived that they achieved differently with respect to the attributes. This finding was somewhat expected, as the students may hold different views and priorities on the attributes presented to them in the questionnaire.

Interaction between stages and perceived benefit and stages and discipline groups

The investigation of the interaction between stages and perceived benefit and stages and discipline groups was conducted only on those attributes that had statistically significant values, as presented in the previous section (Table 4.1, columns

9 and 10). The results show that there is interaction between stages and perceived benefit in three attributes: 'Ability to apply mathematical, scientific and technological tools', 'Ability to use information technology effectively' and 'Analytical in the formulation and solutions of problems'. An illustration of this interaction for 'Ability to use Information Technology effectively' is presented in Figure 4.1. The figure includes two crossing lines, one representing the students who believe the benefit of collaborative learning shows the increase of average score from pre- to post-implementation and the other one representing the students who do not believe the benefit shows the decrease of average score. This finding suggests that those who believe that working on a collaborative design project is going to improve these attributes are more likely to have improved perceived achievement to these attributes; those who do not believe similarly are more likely to not perceive any achievement (or less achievement). Thus, this finding calls for the need to reinforce learners' beliefs in the benefit and self-efficacy early in the implementation.

Despite significant differences in all attributes, the interaction between stages and discipline groups is only found in three attributes: 'Ability to communicate effectively, using both written and oral methods', 'Ability to use information technology effectively' and 'Self-motivated'. An illustration for 'Ability to use information technology effectively' in Figure 4.2 demonstrates four lines with different

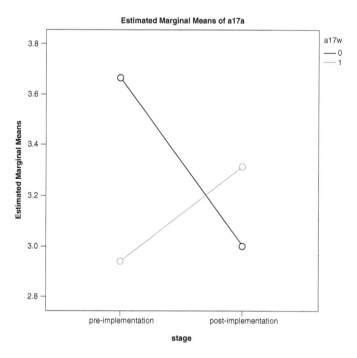

Figure 4.1 An illustration of the interaction between stages and perceived benefit for 'Ability to use information technology effectively'

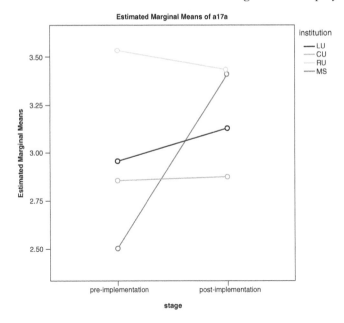

Figure 4.2 An illustration of the interaction between stages and discipline groups for 'Ability to use information technology effectively'

inclination for the pre- and post-implementation scores for different disciplines. While the lines, representing civil/structural engineers (CU), construction managers (LU) and postgraduate sustainability managers (MS) do demonstrate increasing scores in post-implementation, the line for architects (RU) shows decreasing scores. In this particular case, perhaps, the architects did not feel that they had learnt any new software, as it was realised during the implementation they already had sufficient skills to use drawing software to the advantage of their colleagues in other institutions. Nevertheless, this finding suggests that the changes (increase or decrease) in scores are not the same for all discipline groups. One discipline group may not perceive the same benefit another group does.

Attributes grouping and their contribution to successful collaborative design project

The PCA has neatly grouped the attributes into four components which are interrelated (due to the use of oblique rotation; see Table 4.2). The first component includes six attributes related to communication, teamwork and management skills both for team and personal. Self-motivation was also part of this component, and therefore the component is named 'teamwork, management and motivation'. The variance of the first component contributes 43% of the total variance. The fact that its membership covers four of eight most improved attributes confirms its importance. However, the analysis is not meant to be predictive of the improved performance.

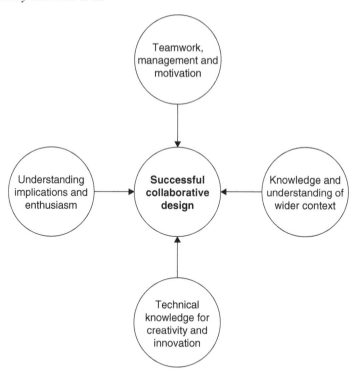

Figure 4.3 Components of required attributes for successful collaborative design

The second component is the largest group, with 10 attributes. It covers a range of attributes which suggest a sound underpinning of basic knowledge and ability to perform a meaningful contribution to the team. This component is called 'technical knowledge for creativity and innovation' and contributes 6.3% of the total variance. The third component is formed on five attributes, mostly related to ability to assess the implications of decisions/actions. An attribute of enthusiasm is also part of the component, which is consequently titled 'understanding implications and enthusiasm'. This component contributes 5% of the total variance. Four attributes related to understanding operational context are subsumed under the fourth component, which is named 'knowledge and understanding of wider context'. This component contributes 4.5% of the total variance.

This finding (as illustrated in Figure 4.3) suggests that for a successful collaborative design project, participants should possess sound technical knowledge, applied using good teamwork skills, understanding direct and wider implications of design decisions and actions with the team.

Conclusions

A collaborative design project can benefit learners in BE subjects in several ways: (i) it trains the students to work in multidisciplinary, collaborative teams, which is conditioned to mimic real working life; (ii) it equips the students with the skills set required for virtual/online collaborative working within globally distributed teams; (iii) it equips the students with 'soft' people management skills, which are believed to provide a competitive advantage in the current job market. The chapter has presented an evaluation of employability attributes measured pre- and post-implementation of a collaborative design project involving students from three international institutions. Although the idea of collaborative design projects in BE education is not new, a comprehensive evaluation of the impact on employ-ability attributes is surprisingly scant. The conclusions are addressed under three research questions:

What are the impacts of a collaborative design project on learning outcomes, defined by attributes including understanding, ability, skills and qualities, as perceived by the participants, pre- and post-implementation?

This study has revealed the positive impact of a collaborative design project with perceived improvement to all attributes (but one) following implementation. The extent of improvement varies across the attributes, with most improved attributes demonstrating a statistical difference between pre- and post-implementation. 'Abil-ity to use information technology effectively' being the most improved attribute is rather surprising. IT-mediated collaboration in a design project was new for most students as they learnt to use IT tools (e.g. GoToMeeting) for communication and discussed issues within a design software environment (e.g. Autodesk Revit). Hav-ing a different level of software skills can have an adverse impact on the collabora-tion experience, as teams may rely on those (most likely architectural) students who have better skills to operate design software (Becerik-Gerber, Ku and Jazizadeh, 2012). In the project under consideration, this issue created imbalance in work load, raising a concern amongst participating students. In most situations, this issue is inevitable given different skills of participants in the real world, but tutors should prepare students to deal with this issue and highlight the potential opportunity from learning new software skills. In any case, it is wise to equip students with basic software skills so that they can collaborate meaningfully online and in real time.

Is there any difference in perceived impact between participants from different disciplines and those with different beliefs about the efficacy of the learning approach?

The analysis indicates the influence of self-efficacy belief on several attributes with significant improved attributes for those who believed that participating in the design project would improve their attributes. This highlights the need for tutors to nurture confidence and self-efficacy and communicate the benefit con-tinuously from the start of implementation. The tutors have an important role in making positive self-efficacy belief habitual as early as possible in the learn-ing endeavour. In group learning, peer achievement is often used as a reference to compare one's own performance with possible positive and negative impact on self-efficacy belief. Here, the tutors should emphasise that individual members of the group should support and complement (rather than compete with) one another's roles. Group membership should be determined based on complementary attributes so that individual members have the opportunities to contribute meaningfully to the group. The tutors can also set proximal goals in conjunction with distal goals, allowing for frequent constructive feedback on individual performance (Pajares, 2002).

The interaction between stages and discipline groups is only found in three attributes. One of them is 'Ability to use information technology effectively', which was attributed to different software skills of the participating students. Considering the low number of attributes which demonstrated interaction, there is a need for further research in this area.

What attributes can facilitate successful collaborative design?

Although the PCA is not meant to be predictive or indicative of the weight of each component, the result suggests four components of required attributes for successful collaborative design. The component 'teamwork, management and motivation' is critical for successful collaborative design, given that the majority of attributes belonging to this component were the most improved attributes as perceived by participating students. The components do not only facilitate successful collaborative design but also encourage and develop employability attributes among the participants.

References

Ayres, R. (2015). Lecturing, Working with Groups and Providing Individual Support, in Fry, H., Ketteridge, S., and Marshall, S. (eds.), *A Handbook for Teaching and Learning in Higher Education*. Oxon: Routledge.

Bandura, A. (1977). Self-Efficacy: Toward a Unifying Theory of Behaviour Change. *Psychological Review*, 84, 191–215.

Beanland, D., and Hadgraft, R. (2014). *Engineering Education: Transformation and Innovation. A Monograph Commissioned by UNESCO*. Melbourne: RMIT University Press.

Becerik-Gerber, B., Ku, K., and Jazizadeh, F. (2012). BIM-Enabled Virtual and Collaborative Construction Engineering and Management. *Journal of Professional Issues in Engineering Education and Practice*, 138(3), 234–245.

Bhandari, A., Ong, S.K., and Steward, B.L. (2011). Student Learning in a Multidisciplinary Sustainable Engineering Course. *Journal of Professional Issues in Engineering Education and Practice*, 137(2), 86–93.

Biggs, J.B. (1996). Enhancing Teaching through Constructive Alignment. *Higher Education*, 32, 1–18.

BIM2050 group. (2014). *Built Environment 2015: A Report on our Digital Future*. London: Construction Industry Council.

CACB. (2012). *Conditions and Terms for Accreditation for Professional Degree Programs in Architecture*. Ottawa, Canada: The Canadian Architectural Certification Board.

CIOB. (2013). *The Education Framework for Undergraduate Programmes: The Chartered Institute of Building*. Berkshire: Chartered Institute of Building.

Egan, J. (2004). *Skills for Sustainable Communities: Office of the Deputy Prime Minister.* London: Office of the Deputy Prime Minister.

Howard, P. (2014). Assessment in PBL Environment, in Beanland, D. and Hadgraft, R. (ed.), *Engineering Education: Transformation and Innovation: A Monograph Commissioned by UNESCO* (pp. 101–104). Melbourne: RMIT University Press.

Korkmaz, S. (2012). Case-Based and Collaborative-Learning Techniques to Teach Delivery of Sustainable Buildings. *Journal of Professional Issues in Engineering Education and Practice*, 138(2), 139–144.

Leinonen, P., Järvelä, S., and Häkkinen, P. (2005). Conceptualizing the Awareness of Collaboration: A Qualitative Study of a Global Virtual Team. *Computer Supported Cooperative Work*, 14, 301–322.

Maddocks, A.P., Dickens, J.G., and Crawford, A.R. (2002). *Encouraging Lifelong Learning by Means of a Web-Based Personal and Professional Development Tool*. Manchester: ICEE, UMIST, 18–22 August.

Male, S.A., Bush, M.B., and Chapman, E.S. (2011). An Australian Study of Generic Competencies Required by Engineers. *European Journal of Engineering Education*, 36(2), 151–163.

O'Brien, W., Soibelman, L., and Elvin, G. (2003). Collaborative Design Processes: An Active- and Reflective-learning Course in Multidisciplinary Collaboration. *Journal of Construction Education*, 8(2), 78–93.

Pajares, F. (2002). Gender and Perceived Self-Efficacy in Self-Regulated Learning. *Theory into Practice*, 41(2), 116–125.

RIBA. (2014). *RIBA Procedures for Validation and Validation Criteria for UK and International Courses and Examinations in Architecture*. London: Royal Institute of British Architects.

Soetanto, R., Childs, M., Poh, P., Austin, S., and Hao, J. (2014). Virtual Collaborative Learning for Building Design: Proceedings of the Institution of Civil Engineers – Management. *Procurement and Law*, 167, MP1, 25–34. http://dx.doi.org/10.1680/mpal.13.00002.

Solnosky, R., Parfitt, M.K., and Holland, R. (2014). IPD and BIM-Focused Capstone Course Based on the AEC Industry Needs and Involvement. *Journal of Professional Issues in Engineering Education and Practice, Special Issue: Curriculum Assessment and Continuous Improvement*, 140(4). doi: 10.1061/(ASCE)EI.1943–5541.0000157.

Solnosky, R., Parfitt, M.K., and Holland, R. (2014). Delivery Methods for a Multi-Disciplinary Architectural Engineering Capstone Design Course. *Architectural Engineering and Design Management*, doi: 10.1080/17452007.2014.925418.

Stanford, M.S., Benson, L.C., Alluri, P., Martin, W.D., Klotz, L.E., Ogle, J.H., Kaye, N., Sarasua, W., and Schiff, S. (2013). Evaluating Student and Faculty Outcomes for a Real-world Capstone Project with Sustainability Considerations. *Journal of Professional Issues in Engineering Education and Practice*, 139(2), 123–133.

Tlhoaele, M., Suhre, C., and Hofman, A. (2015). Using Technology-Enhanced, Cooperative, Group-Project Learning for Student Comprehension and Academic Performance. *European Journal of Engineering Education*, doi: 10.1080/03043797.2015.1056102.

Tucker, R., and Rollo, J. (2006). Teaching and Learning in Collaborative Group Design Projects. *Architectural Engineering and Design Management*, 2, 19–30.

Tymon, A. (2013). The Student Perspective on Employability. *Studies in Higher Education*, 38(6), 841–856.

Vygotsky, L.S. (1978). *Mind in Society: The Development of Higher Psychological Processes*. Cambridge, MA: Harvard University Press.

Wolcott, M., Brown, S., King, M., Ascher-Barnstone, D., Beyreuther, T., and Olsen, K. (2011). Model for Faculty, Student, and Practitioner Development in Sustainability Engineering through an Integrated Design Experience. *Journal of Professional Issues in Engineering Education and Practice*, 137(2), 94–101.

Zimmerman, B.J. (2000). Self-Efficacy: An Essential Motive to Learn. *Contemporary Educational Psychology*, 25, 82–91.

5 BIM-Hub project evaluation

Principles, protocols, methodologies and outcomes

Harry Tolley and Helen Mackenzie

> Building Information Modelling (BIM) is a collaborative way of working, under-pinned by the digital technologies which unlock more efficient methods of design-ing, creating and maintaining our assets.
>
> (BIS, 2012)

Introduction

The starting point for developing an evaluation strategy for the BIM-Hub project was an agreement that its primary purpose should be to collect and analyse data, the outcomes of which would be fed into debates held amongst stakeholders with a view to informing decision making throughout its life history. In practice this meant that evaluation was deeply embedded in the implementation of the project's activities, featured prominently in its strategic and operational plans, was dis-cussed regularly at consultative and steering group meetings and was conducted continuously throughout the life of the project – and not as an afterthought that was 'bolted on' as it approached its conclusion.

Guidelines written to support the implementation of earlier funding initiatives in the UK aimed at enhancing the quality of teaching and learning in HE were helpful in terms of identifying some of the key principles on which the evaluation of the project should be based, for example by the Higher Education Employ-ment Division (1998a and 1998b), the Higher Education Funding Councils (1999; Peters (2007) and the National HE STEM Programme (Moore, 2011). However, the complex nature of the curriculum developments that were to be undertaken by the BIM-Hub project (and the fact that the digital representations and manage-ment of the physical and functional characteristics of buildings was central to the planned learning activities) prompted the project to draw upon other guid-ance, particularly that offered by the Joint Information Systems Committee (JISC) through Glenaffric Ltd. (2007). The six actions set out in those guidelines sug-gested that the evaluation of a project should begin with the identification of its key stakeholders and their social networks and an analysis of their interests and involvement in the project in terms of their particular roles and responsibilities. This was to be followed by an attempt to describe and understand the project in terms of its intended outcomes, methods of delivery (including the application of

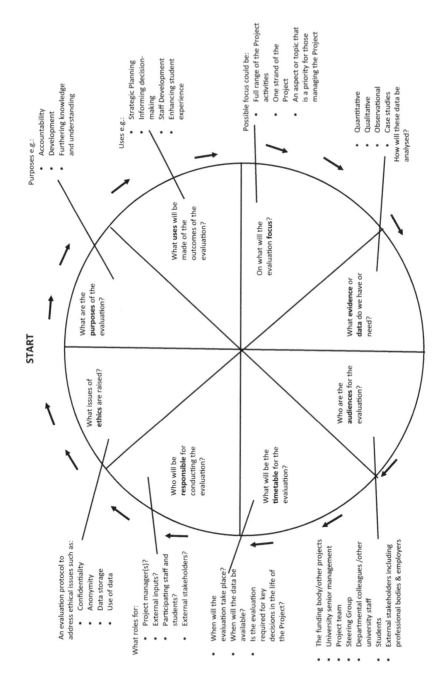

Figure 5.1 Developing a strategic plan for evaluation

ideas drawn from the literature relating to teaching and learning), timescale, institutional context and place within the wider development programme supported by the external funding body.

In other words, a thorough situational analysis was seen as a necessary precursor to the next step in the process (i.e. designing the evaluation), which would culminate in the development of a plan that addressed a series of practical questions (see Figure 5.1) and ensure that it was constructively aligned with the project's strategic and operational plans. In so doing it articulated a rationale for the evaluation, identified the methods of data collection and analysis to be used in relation to the intended pedagogy, set out a timescale for the implementation of those activities closely linked to the project's operational plan, defined the roles and responsibilities of different stakeholders and indicated how and to whom feedback from the evaluation would be reported and discussed. The plan also addressed the need for all aspects of the evaluation to be undertaken within an ethical framework such as those advocated by the British Educational Research Association (BERA, 2011) and the UK Evaluation Society's Good Practice Guidelines (www. evaluation.org.uk/).[1]

Purposes served by evaluation

Chelimsky's (1997) distinction between what she called evaluation *'perspectives'*, and which Saunders (2000) called *'uses'*, was also seen as a useful starting point for discussion about the formulation and implementation of the project's evaluation plan. The three main uses these writers identified were evaluation for the purposes of:

- Accountability, for example judging outcomes, effectiveness, efficiency and success as the basis for monitoring progress against the project's strategic and operational plans, summative evaluation and interim and final reporting.
- Development, such as by providing formative feedback aimed at improving the design and delivery of students' learning experiences, furthering the continuous professional development (CPD) of academic staff and improving the project's ongoing decision making.
- Furthering educational knowledge by collecting and analysing data, which would lead to an enhanced understanding of specific pedagogic activities, aspects of HE policy and the planning and management of curriculum change.

It was apparent, therefore, that overlaps would exist between evaluation undertaken for the purposes of furthering development and action research and evaluation aimed at advancing knowledge and understanding and educational or pedagogic research. The **aim** of the BIB-Hub project was that its evaluation would address all three of the listed purposes and do so through a strategic plan based on the adoption of action research methodology, implemented by means of two cycles of developmental activity of equal length across the BIM-Hub project and its preceding project. Such an approach was consistent with the proposals set out on the BIM mode of working (**see** bim-hub.lboro.ac.uk/guidance-notes/level-5/).

Subjects or stakeholders

According to the JISC guidelines, however, the first step to effective evaluation was to identify the stakeholders and analyse their particular interests in the implementation of the project and its outcomes. Clearly, that list included all those who would be directly involved in the design and delivery of the project and affected by its outcomes, such as student groups and academic staff from the three participating universities,[2] the manager, researcher and external evaluators. Importantly, the students were seen from the outset not as *'consumers'* and providers of feedback on their experience of a curriculum innovation that had been designed for them by others but as *'co-creators'* in the generation of a shared development activity (see e.g. McCulloch, 2009; NUS, 2009). This was consistent with the adoption of action research methodology, which according to Punch (1998) involves *'subjects'* (e.g. students and academic staff) being seen as equal partners in the research process, that is as 'respondents, participants, and stakeholders' (p. 169) and warning that 'If evaluators cannot be clear, direct and undeceptive regarding their wish to know how stakeholders make sense of their contexts, then stakeholders will be unclear, indirect and probably misleading regarding how they too engage in sense-making and what their basic values are' (Guba and Lincoln, 1989:22 cited in Punch, 1988; 170). Consequently, students were represented (alongside their academic tutors) at all of the meetings[3] of the project's consultative committee and steering group at which there were specific items on the agenda (such as evaluation) on which they were expected to guide the discussion by providing feedback from the learners' perspective on the issues under consideration. Beyond those listed with a direct and on-going involvement in the project it was recognised that there were others with a legitimate stake in its progress and outcomes including the funding body (i.e. the Higher Education Academy), the organisations (e.g. the Institution of Civil Engineers or ICE) that validate and accredit the relevant degree programmes in the participating universities and the wider communities of interested HE academics and professional building industry practitioners.

BIM-Hub project as action research

The timescale of the BIM-Hub project's external funding enabled an operational plan to be developed that was based on the implementation of two action research cycles involving the same academic staff, researcher and external evaluators but with two different sets of students working in groups made up of final-year students from the three participating universities. The intended learning outcomes, methods of delivery and modes of assessment in both cycles would be broadly the same but with some adjustments to the second in the light of evaluative feedback from the first. The approach that was adopted was consistent with the ideas of Clegg (2000) on 'knowing through reflective practice' in HE, Kemmis (1982) of action research as 'self-reflective enquiry', Allwright and Bailey (1991) as 'exploratory teaching and learning' and Kemmis and McTaggart (1992) that compared with normal practice it 'is more systematic and collaborative in collecting evidence on which to base their group reflection'. It was motivated, therefore, by a commitment to function as a self-critical community with a view to improving the design and delivery of the HE curriculum; furthering the professional development of its academic staff; and providing the valid and reliable evidence on which future decision-making could be based, and the outcomes of the project could be reported for the purposes of accountability and disseminated to a wider audience via its website, conference presentations and publications. The guidance on the design and implementation of action research methodology offered by Carr and Kemmis (1986), Cohen and Manion (1989), Kemmis (1997) and Cohen, Manion and Morrison (2000) was followed to further the achievement of those outcomes.

Integrating BIM into the HE curriculum

The development activities undertaken by the BIM-Hub project centred on the design and delivery of a module located in the degree programmes offered at each of the participating universities. The intended learning outcomes of that unit of study, its mode of delivery, methods of assessment and timetabling were agreed and coordinated across all three institutions, care being taken to ensure that it complied with the accreditation requirements of the relevant professional bodies. Its underlying **aim** was to further the students' employability through the experience of working over a period of time in groups made up of final-year students from each of the three universities. In the groups to which they were allocated, the students were expected to work together as a team and using their subject specialisms (e.g. in architecture, construction management and civil engineering) to devise practical solutions to authentic design problems associated with the proposed development of an urban building site and, in so doing, use BIM modes of working.

In keeping with the industrial practices they were being asked to simulate, the groups were given a specification for the completion of their design project and instructed to take responsibility for convening a number of virtual team meetings in order to manage and advance that work. The intention was that this would give the students authentic reasons for using the GoToMeeting videoconferencing platform and, in so doing, gain experience of synchronous decision-making based on the critical examination by the group of calculations, diagrams and plans that they themselves had produced. It was anticipated that the adoption of this BIM mode of working would enable the knowledge, skills and personal attributes held within the groups to be used constructively for the purposes of resolving the building design problems presented in the simulation and that in the process of collaborating with each other new learning would be socially constructed. Of particular significance in this context was the opportunity this experience offered for the development of transferable skills in communication, recognising and responding appropriately to cultural and sub-cultural differences[4] in teamworking and for the acquisition of what is known as *'shared situation awareness'*. Such a sensibility is known to be especially important in working environments involving collaboration in teams where the information flow is high and from culturally diverse sources. In such situations, failure to distinguish what is important from what is not, misinterpretation of what has been said (or not said) and insensitivity to the significance of silence and to the looks and gestures that accompany speech can lead to poor and ill-informed decision-making (see e.g. Endsley, 2004).

The pedagogy devised for the purposes of integrating BIM into the HE curriculum, therefore, differed significantly in both theory and practice from the problem-based learning (PBL) developed elsewhere (e.g. in relation to medical education as described by Colliver [2000] and Davies *et al.* [1999]) in which a problem is given to groups of students as the starting point for their learning. In such PBL models, the students acquire new subject knowledge and transferable skills through solving a staged sequence of problems presented to them in context, together with the associated learning materials and on-going access to tutors acting as facilitators of their learning. When the module was first being planned (in 2014) it would have been difficult to have adopted such a structured approach for the induction of students into BIM modes of working – not least because in this regard new digital technologies and related professional practices were still in the process of being developed, documented and exemplified ahead of the introduction of building industry compliance in 2016 (BIS, 2012; op cit).

In the circumstances it was decided that the intended learning outcomes of the module (and the wider objectives of the BIM-Hub project) could be achieved through the engagement of the students in a form of *'academic play'* enacted in a *'transitional space'* within which they would be invited to tackle carefully selected but nevertheless challenging versions of authentic building design problems and, in doing so, gain experience of the BIM mode of working. The ideas of academic play and the concept of transitional space as discussed here are based upon the seminal ideas of the British psychoanalyst Donald Winnicott (1896–1971). Winnicott's work is now beginning to be adopted more widely by writers and scholars within a wide range of disciplines (see Kuhn, 2013), but it has been largely overlooked in the context of HE. There are, however, some notable exceptions including Creme and Hunt (2002), Creme (2008) and Mackenzie (2011).[5] Whilst a detailed discussion of the underlying concepts falls outside the scope of this chapter, a brief summary of their application to teaching and learning in HE is given in what follows in order to provide insights into both the nature of the students' learning experiences and the challenges this posed for those seeking to evaluate them for the purposes outlined.

Pedagogy based on the principles of academic play gives students an opportunity to experience:

- **Personal Development:** Winnicott's overarching theory of emotional development focuses upon the *'evolution of personality and character'* (Winnicott, 1958; 3) and thus a sense of *'self'* and *'identity'*. Whilst Winnicott's work was with children, he acknowledged that personal development occurs throughout life and that everyone is continuously engaged in the process of becoming someone.
- **Creative Play:** Winnicott (1971) saw *'creative, physical, and mental activity'* (p. 75) as play that forms the basis for creative living and cultural experiences in adult life. It is through play that it is possible for individuals to search for and discover the *'self'* (ibid.) and thus develop their identities.[6] Winnicott (1971; 148) proposed that a 'third area of human living' (or *'transitional space'*) must be opened up in order to achieve such experiences.
- **Transitional Space:** According to Winnicott (ibid, p. 138) transitional space exists between the *'inner or personal psychic reality'* and the outer *'actual world'* in which the individual lives. Winnicott (ibid) referred to this as the *'intermediate zone'* (p. 141) because it is positioned between the *'inner'* self and *'outer'* world. He also saw this as a *'potential space'* (p. 55) because it is the individual who must create it and make it exist since there is nothing that makes the space inevitably transitional. Transitional space, therefore, involves active engagement in creative play and cultural experiences in which the focus is upon personal development involving freedom from direction and the gradual growth of individual autonomy.
- **Freedom:** Winnicott suggested that individuals move from dependence on others towards becoming more autonomous through creative play within a transitional space aided by the provision of a facilitative learning environment created specifically for the purpose.
- **A Facilitative Holding Environment:**[7] Environmental provision plays a key role in facilitating the difficult transitions that individuals face (Winnicott, 1971) by helping them to remain sustained in transitional space whilst developing a strengthening sense of self. Ideally, that provision is designed to be *'just good enough'*, that is not too structured and over protective, and aided by the presence of transitional objects and phenomena (Winnicott, 1951) that serve as a bridge between the familiar and the disturbingly unfamiliar. This suggests that Winnicott did not view the experience of transition as smooth and linear but rather as one that might prove to be recursive, difficult, challenging and possibly risky.

The learning activities delivered to students over the course of the two action research cycles implemented by the BIM-Hub project can be viewed as *'academic play'* in that they were designed to provide them with a substantial degree of freedom to direct and manage their own learning whilst working within a facilitative environment deliberately created for them by their academic tutors. Whilst much of that learning occurred whilst the students were actively engaged with others in relation to furthering the completion of the common tasks (i.e. in team meetings at which building design issues were discussed and resolved), it was also achieved through individual research, creating new building designs, devising solutions to structural engineering problems, informal discussion and personal reflection that occurred at other times and in different locations. In addition to attending and contributing to the meeting of their teams, the students were required to sub-mit individual reports at the end of the module for assessment by their academic tutors.[8] That report included an element of personal reflection on what they had contributed and learned from the experience of the module as a whole – including their collaboration with others in their working groups. They were also asked to submit a peer assessment (using the structured guidance in the form of criteria provided by an on-line tool known as WebPAvn[9]) of the contribution each student had made to the project work undertaken by their group. The module, therefore, in the way that it was created and delivered, offered students an opportunity to develop an enhanced awareness of their own identity (or 'self') as they moved towards becoming qualified professionals in their chosen fields of study.

Academic tutors from the three universities were responsible for creating that *'facilitative holding environment'* for their students, that is they produced the working brief which defined the design problem to be tackled, articulated the intended learning outcomes and set out the institutional requirements for the pro-ject with regard to the overall management of the module, the procedures to be followed, the assessment requirements and the deadlines to be met. Those tutors were not present in person, however, to facilitate the learning process when the groups held their virtual meetings. In their absence each group was free to plan and manage its own use of time, arrange and conduct its meetings and resolve any technical, communication and teamworking problems it encountered. Neverthe-less, tutors had a presence in that transitional space in that should they be needed they were available for consultation by the students outside of the team meetings. As authority figures in their subjects they were seen as potential sources of expert knowledge who could be called upon if needed by the students to give advice on technical problems encountered by the groups – and if necessary clarify the requirements of the specification the students had been given as an integral part of the facilitative holding environment created for them.

Evaluating the BIM-Hub project: challenges and methods

As many writers have explained (e.g. Baume, 2003, 2005) evaluating curriculum innovation at any level of education for purposes such as those outlined is sel-dom a simple matter, not least because of the difficulties involved in determining outcomes and distinguishing causes and their effects. The evaluation of the BIM-Hub project was no exception for a variety of reasons, including the requirements to address the needs of different stakeholders for both formative and summative feedback, monitor the processes by which two action research cycles were planned and implemented and disseminate the outcomes to multiple audiences. The major challenges, however, centred on evaluating within the limited timescale of the project its impact on the learning experiences students had acquired through working in groups on a building design project using BIM and the continuing professional development of their academic tutors. Unlike most curriculum development initiatives, educational change is a long-term process – many of the intended outcomes taking some time before they are achieved, often only becoming evident long after a project and its related evaluation activities have ceased to function. This is particularly significant in relation to student learning acquired through experiences such as those provided by the BIM-Hub project. Many of the learning outcomes the students were intended to acquire with a view to enhancing their future employability (e.g. in problem solving, time management, team working, intercultural communication, synchronous decision making and self-reliance), along with a growing sense of becoming graduate building industry professionals, may not have become fully evident in the short term – either to their tutors observing their behaviour or to themselves. Indeed, the newly acquired knowledge and skills relevant to the workplace may well remain tacit and unnoticed for some time, perhaps residing in the individual at the level of unconscious competence despite efforts to offset that through planned interventions such as prompting students to reflect on their learning experiences when writing their reports, conducting peer assessments using WebPA, being interviewed and participating in focus groups.

It was evident, therefore, that in order to address these challenges, evidence would need to be gathered from different perspectives, that multiple methods would have to be used in the collection and analysis of that data and that the process would need to be both theoretically informed and rigorously implemented. To that end, use was made of quantitative data collected in all three institutions by means of end-of-module questionnaires of student satisfaction with regard to their learning experiences. The survey results were made available for use in routine teaching quality assurance procedures and also helped to inform subsequent data collection using qualitative methods. Semi-structured interviews (Drever, 1995; Kvale and Brinkman, 2009) were conducted with a small sample of students from all three universities after the completion on both occasions of the teaching module and with their academic tutors at the end of the first action research cycle. Issues that emerged from the questionnaire survey and the interviews were explored further with samples of students in all three universities by means of focus groups (Krueger, 1988; Morgan, 1988). In addition, the students' reflections on their experience over the course of the module, which they had captured in 1,000 words as part of their final reports, were subjected to textual analysis. In this context, reflection was defined as the conscious analysis of their experiences with a view to identifying lessons from them for future use (see e.g. Atkins & Murphy, 1994; Boud and Walker, 1998; Hogston and Simpson, 1999). Finally, each team submitted an audio-visual recording of one of its virtual meetings at which GoToMeeting had been used. These presented an opportunity to review the processes by which the meetings had been transacted including how synchronous decisions were reached and the way BIM was being used and where its potential remained unexploited. When triangulated with evidence derived from other sources the outcomes of this analysis were used in the production of *'Guidance notes for supporting online collaboration for design'* (see: http://bim-hub. lboro.ac.uk/guidance-notes/).

Conclusions

With regard to the evaluation of curriculum innovation in HE, experience of the BIM-Hub project highlighted the importance of:

- Establishing clear goals for planned changes to the HE curriculum (and for the related monitoring and evaluation) in order to ensure that valid and reliable evidence is collected that can then be used to inform future decision making.
- Embedding evaluation into the planning and delivery of the project rather than bolting it on at the end.
- Being alert to the possible occurrence of unexpected (or serendipitous) outcomes and prepared to recognise their potential value in the quest for continuous improvement.
- Identifying the key stakeholders in the project and the nature of their interests in the proposed innovations.
- Negotiating and agreeing with those stakeholders the purposes of the evaluation, the methods by which it will be delivered and their roles in its implementation.
- Ensuring that the evaluation is conducted systematically according to agreed strategic and operational plans that are closely aligned with those for the project as a whole.
- Making certain that the evaluation is undertaken in a spirit of collaboration between all stakeholders acting as a self-critical community.
- Conducting all aspects of the evaluation according to an agreed ethical framework with regard to the collection, analysis, storage and use of data.
- Using all available means of communication (e.g., text messages, emails, face-to-face meetings and Skype) to ensure that those responsible for the evaluation of the project are kept up to date with its progression, including matters giving cause for concern to any of the stakeholders.
- Making sure that the evaluation is scholarly and theoretically informed through references to the relevant literature.
- Employing a range of research methods (both quantitative and qualitative) for collecting and analysing evaluation data.
- Subjecting evidence derived from different perspectives (and obtained through the use of alternative methods) to a rigorous process of triangulation.
- Acknowledging the limitations of evaluation as an activity in terms of what it can and cannot successfully capture.
- Recognising that many of the project's outcomes may not become evident (and hence available for collection and analysis) until some time after the end of the project and its evaluation.

- Appointing external evaluators (and researchers) who are capable of exercising independent judgements, acting as 'critical friends' to the project team and providing it with access to information and ideas derived from the literature.

Notes

1 This matter was discussed at initial briefing meetings with the students, who were invited to sign an agreement/consent form about the protocols that would be followed for the ethical collection, storage and use of evaluation data, including the rights of participants to confidentiality and anonymity.
2 Loughborough University (UK), Coventry University (UK) and Ryerson University (Ontario, Canada).
3 These were held at regular intervals by means of videoconferencing using GoToMeeting.
4 Intercultural communication is increasingly seen as a matter of great importance, not just because of increased globalisation of business activity but also because different age groups, genders, subject disciplines and professions have their own implicit knowledge, rules, beliefs and values.
5 Also relevant to HE is the writing of Ellsworth (2005), who uses Winnicott's ideas, alongside others, to explore different learning places, focusing upon media, architecture and pedagogy.
6 Creme (2008) has used these ideas in a HE context to develop the notion of 'academic play'.
7 A 'holding' environment involves ideas around the provision of support and comfort to facilitate transition.
8 These could be regarded as familiar *'transitional objects'* that acted as a bridge for the students between the familiar and the disturbingly unfamiliar.
9 WebPA is maintained by the Centre for Engineering and Design Education (CEDE) at Loughborough University.

Bibliography

Allwright, D., and Bailey, K.M. (1991). *Focus on the Language Classroom: An Introduction to Classroom Research for Language Teachers*. Cambridge: Cambridge University Press (CUP).

Atkins, S., and Murphy, K. (1994). Reflective Practice. *Nursing Standard*, 8(39), 49–56.

Baume, D. (2003). Monitoring and Evaluating Staff and Educational Development, in Kahn P. and Baume, D. (eds.), *A Guide to Staff and Educational Development* (pp. 63–78). London: Kogan Page.

Baume, D. (2005). Monitoring and Evaluating Employability Ventures, in *Enhancing Student Employability: Higher Education and Workforce Development*. Birmingham, Quality Research International.

BIS. (2012). Industrial Strategy: Government and Industry in Partnership: Building Information Modelling. www.bis.gov.uk/assets/BISCore/economics-and-statistics/docs/I/12–1140-industrial-strategy-uk-sector-analysis.pdf. [Accessed 17th February 2016].

BIS. (2015). Fulfilling Our Potential: Teaching Excellence, Social Mobility and Student Choice. www.gov.uk/. . ./BIS-15–623-fulfilling-our-potential-teaching-excecellence. [Accessed 17th February 2016].

Boud, D.J., and Walker, D. (1998). Promoting Reflection in Professional Courses: The Challenge of Context. *Studies in Higher Education*, 23(2), 191–206.

British Educational Research Association. (2011). Ethical Guidelines for Educational Research. www.bera.ac.uk/wp-content/.../2014/02/BERA-Ethical-Guidelines-2011-pdf. [Accessed 2nd January 2014].

Carr, W., and Kemmis, S. (1986). *Becoming Critical: Education, Knowledge and Action Research*. Lewes: Falmer Press.

Chelimsky, E. (1997). Thoughts for a New Evaluation Society. *Evaluation*, 3(1), 97–109.

Clegg, S. (2000). Knowing through Reflective Practice in Higher Education. *Educational Action Research*, 8(3), 451–469.

Cohen, L., and Manion, L. (1989). *Research Methods in Education*. London: Routledge.

Cohen, L., Manion, L., and Morrison, K. (2000). *Research Methods in Education*. London: Routledge.

Colliver, J.A. (2000). Effectiveness of Problem-Based Learning Curricula: Research and Theory. *Academic Medicine*, 75(3), 259–266.

Creme, P. (2008). A Space for Academic Play: Student Learning Journals as Transitional Writing. *Arts and Humanities in Higher Education*, 7(1), 49–64.

Creme, P., and Hunt, C. (2002). Creative Participation in the Essay Writing Process. *Arts and Humanities in Higher Education*, 1(2), 145–166.

Davies, D., Thomson O'Brien, M.A., Freemantle, N., Wolf, F.M., Mazmanian, P., and Taylor-Vaisey, A. (1999). Impact of Formal Continuing Medical Education, J.A.M.A., 282 (9), 867–874.

Drever, E. (1995). *Using Semi–Structured Interviews in Small–Scale Research: A Teacher's Guide*. Edinburgh: Scottish Council for Research in Education.

Ellsworth, E. (2005). *Places of Learning: Media Architecture Pedagogy*. New York & London: Routledge Falmer.

Endsley, M.R. (2004). Situation Awareness: Progress and Directions, in Banbury, S. and Tremblay, S. (eds.), *A Cognitive Approach to Situation Awareness: Theory and Application* (pp. 317–341) Aldershot: Ashgate Publishing.

Glenaffric Ltd. (2007). *Six Steps to Effective Evaluation: A handbook for programme and project managers, and Evaluation Framework*. Bristol: Joint Information Systems Committee (JISC).

HEED. (1998a). Evaluating Development in Higher Education: A Guide for Contractors and Project Staff, Higher Education Employment Division, Department for Education and Employment. Moorfoot, Sheffield: DfEE.

HEED. (1998b). *Evaluating Development in Higher Education: A Guide for Steering Group Members, Higher Education Employment Division, Department for Education and Employment*. Moorfoot, Sheffield: DfEE.

HEFC. (1999). Monitoring and Evaluation, *Project Briefing Paper No 3*, Fund for the Development of Teaching and Learning (FDTL) and Teaching and Learning Technology Programme (TLTP), Higher Education Funding Councils (HEFC).

Hogston, R., and Simpson, P.M. (1999). *Foundations of Nursing Practice*. London: Palgrave Macmillan.

Kemmis, S. (1982). 'Action Research', in Husen. T and Postlethwaite. T. (ed.), *International Encyclopedia of Education: Research & Studies* (pp. 173–179). Oxford: Pergamon Press.

Kemmis, S. (1997). 'Action Research', in Keeves, J.P. (ed.), *Educational Research Methodology and Measurement: An International Handbook* (pp. 173–179) Oxford: Elsevier Science Ltd.

Kemmis, S., and McTaggart, R. (1982 & 1992). *The Action Research Planner*. Geelong, Victoria: Deakin University Press.

Krueger, R.A. (1988). *Focus Groups: A Practical Guide for Applied Research*. Beverly Hills: Sage Publications.

Kuhn, A (ed.) (2013). *Little Madnesses: Winnicott, Transitional Phenomena and Cultural Experience*. London: I.B. Tauris & Co Ltd.

Kvale, S., and Brinkmann, S. (2009). *Interviews: Learning the Craft of Qualitative Research Interviewing* (Second edition). Los Angeles, London, New Delhi and Singapore: SAGE Publications.

Mackenzie, H.E. (2011). *Students' Experiences of Academic Play within Transitional Space in Higher Education*. Unpublished PhD Thesis: University of Nottingham.

McCulloch, A. (2009). The Student as Co-Producer: Learning from Public Administration about the Student University Relationship. *Studies in Higher Education*, 34(2), 171–183.

Moore, I. (2011). *The National HE STEM Programme Curriculum Innovation Projects. A Guide to Practice: Evaluating your Teaching Innovation*. Birmingham: University of Birmingham. Available at: www.hestem.ac.uk/sites/default/files/evaluatingyourteachinginnovation.pdf. [Accessed 02 January 2016].

Morgan, D.L. (1988). *Focus Groups as Qualitative Research*. Beverly Hills: Sage Publications.

NUS. (2009). *Surfing the Wave. A Strategic Response to a Wave of Change: The Future Landscape of the Student Movement*. London: National Union of Students.

Peters, J. (2007). Using the Personal Development Planning Evaluation Guides and Tools, Centre for Recording Achievement. Available at: www.recordingachievement.ac.uk/images/toolkit/evaluationtoolkit.pdf. [Accessed 10 February 2016].

Punch, M. (1998). Politics and Ethics in Qualitative Research, in Denzin, N.K. and Lincoln, Y.S. (eds.), *The Landscape of Qualitative Research* (pp. 139–164). Thousand Oaks, CA, London & New Delhi: Sage Publications.

Saunders, M (2000). Beginning an Evaluation with RUFDATA: Theorising a Practical Approach to Evaluation Planning, *In Evaluation*, 6(1), 7–21.

Schunk, D.H., and Zimmerman, B.J. (1998). *Self-Regulated Learning: From Teaching to Self-Reflective Practice*. New York & London: The Guilford Press.

Winnicott, D.W. (1951). Transitional Objects and Transitional Phenomena, in Winnicott, D.W. (1984) (ed.), *Through Paediatrics to Psychoanalysis: Collected Papers* (pp. 229–242). London: Karnac.

Winnicott, D.W. (1958). The First Year of Life: Modern Views on the Emotional Development, in Winnicott, D.W. (1965) (ed.), *The Family and Individual Development* (pp. 3–14). London & New York: Routledge.

Winnicott, D.W. (1971). *Playing and Reality*. London & New York: Routledge Classics.

Part II

Complementary case studies

6 A comparison of MOOC development and delivery approaches

Neil Smith, Helen Caldwell, Mike Richards and Arosha K. Bandara

We present a comparison of two ways of developing and delivering massive open online courses (MOOCs). One was developed by the Open University in collaboration with FutureLearn; the other was developed independently by a small team at Northampton University. The different approaches had very different profiles of pedagogic flexibility, cost, development processes, institutional support, and participant numbers. This comparison shows that, several years after MOOCs became prominent, there are many viable approaches for MOOCs. MOOCs on existing large platforms can reach thousands of people but constrain pedagogical choice. Self-made MOOCs have smaller audiences but can target them more effectively.

The range of MOOCs

The MOOC, the massive open online course, has a long history. The MOOC phenomenon builds on a long history of distance education but takes it into the modern online world. Large-scale interaction systems, using technology developed for social networks and e-commerce, have been repurposed to deliver education at a large scale to many students at once. Some of the largest courses have had more than 160,000 students learning concurrently (Hyman, 2012). This potential large reach and the changes it allows in educational providers give MOOCs the potential to foster great innovation in education (Sharples *et al.*, 2014).

However, different MOOCs can use the different aspects of "massive" and "online" in different ways. Moving a course online frees it from the constraints of a physical teaching environment, allowing students to participate in the course without being present in the same place as the teaching staff and often not present at the same time as the teachers. Elements of this have been present in blended learning (Garrison and Kanuka, 2004) courses for several years, where learning activities are moved outside the classroom and students are able to study at the time and place of their choosing, using teaching materials provided, often online.

Since the take-off of MOOCs as a phenomenon in 2012, several companies and universities have started to offer a range of MOOCs. This is in addition to the tools becoming more usable by a wider variety of educators. Together, this increased range of MOOC platforms has led to a wide variety of MOOCs offered to different audiences.

The demand for MOOCs varies widely by size, interest, prior experience, and many other factors. There is therefore a challenge for educators to select the correct pedagogic style of MOOC and the correct delivery style to meet the needs of both the educators and the students.

This chapter outlines the authors' experiences with developing and delivering MOOCs for two very different audiences, with different requirements, and delivered on very different platforms. One MOOC was on cyber security and was a large-scale course for tens of thousands of non-specialist participants. This MOOC was produced by the Open University in collaboration with FutureLearn, a UK-based MOOC delivery company set up with the backing of several UK universities. The other MOOC was on integrating digital tablets (such as iPads) into teaching across a range of subjects and contexts. This was a much smaller course for a few hundred participants and was produced entirely in house by the University of Northampton.

Comparison of MOOCs

As we said, MOOCs can vary in a number of ways. In this section, we outline some of these variations and indicate some criteria that should be used when selecting the most suitable approach when developing a new MOOC.

Audience

MOOCs vary in both their intended and actual audience. The audience can vary in both size and expertise. For instance, the cyber security MOOC was intended for a large and non-specialist audience, giving them some understanding of risks to individuals and some simple techniques to mitigate them. In contrast, the Teaching with Tablets MOOC was intended for in-service educators (in a school, higher education, or further education context).

These different audiences allow MOOC creators to make different assumptions about the interest, commitment, and level of expertise of the participants, and this affects how the MOOC is designed. MOOCs designed for learners with particular skills or in a particular context will necessarily have a smaller potential audience than one for a less particular audience. In addition, the more selective audience could have more commitment to the MOOC; if the learning delivered by the MOOC aligns with their professional or personal interests, they may be more willing to engage in more demanding learning activities over a longer time.

In contrast, MOOCs for a general audience should be carefully designed to reduce barriers to participation for their participants. The "open" nature of the MOOC means that large numbers of people can sign up to MOOCs almost on a whim but then not engage with the MOOC once it starts or drop out before they have completed all the activities. Drop-out rates of more than 90% are common (Khalil and Ebner, 2014), particularly on MOOCs for the general public. But even if an MOOC is designed for a large, general audience, it is another matter to enrol that audience on the MOOC. This is a feature through which the choice of MOOC delivery platform can have a significant effect.

Pedagogy

The first MOOCs (Stacey, 2014), now termed cMOOCs, used a social constructivist pedagogy in which participants developed a shared understanding of the
topic simultaneously with forming a community of practice around the subject,
but these MOOCs are sometimes considered too open ended and woolly (Nkuyubwatsi, 2013). Other MOOCs, termed xMOOCs, have adopted a much more
didactic approach in which students read or watch pre-prepared material and
complete automatically marked exercises. Predictably, xMOOCs have sometimes
been criticised for being too directive.

There is a range of pedagogic approaches between these two extremes, and
there is potential to adopt a nuanced design that navigates these poles in a way that
is appropriate for the audience and subject (Conole, 2013). Again, the pedagogic
approach taken in an MOOC will have a significant impact on the design of the
course.

Platform and services

Choice of platform is not just a technical decision, as different platforms have
different processes embedded within them and can provide different levels of support for MOOC creators. A variety of platforms have grown up for delivering
education online. Many MOOCs, especially those delivered by larger providers
such as Udacity and EdX, use bespoke MOOC web platforms to host all the content and student interaction, as well as provide the back-end services for student
registration, content creation by course authors, and so on. Some MOOCs use
existing virtual learning environment (VLE) platforms to deliver pre-prepared
content, host student-generated content, and provide a forum for discussion. Some
MOOCs, such as the Teaching with Tablets MOOC described in what follows,
assemble a particular student engagement platform from a range of VLE and
social networking platforms used in concert.

Generally, bespoke MOOC platforms are designed for large audiences of
general public as learners. They will often have a single, prescribed pedagogic
approach, generally a didactic approach with readings, video clips, and automatically marked formative assessment tasks. There will generally be some facility
for student interaction through a forum or question–answer tracking system, but
these are often limited in flexibility. Because they are designed for the delivery of
an MOOC to a large general audience, the delivery platform is designed to make
involvement in the course as smooth as possible for the participant.

MOOC platforms provided by large MOOC organisations have other advantages in the support they can provide educators in creating and delivering MOOCs.
As our experience with the cyber security MOOC shows, MOOC providers like
FutureLearn have a robust process for creating and refining MOOCs, including
technical and editorial support for the creation of learning content. They also
tend to have an established base of learners and good publicity mechanisms. This
allows the providers to gather large audiences of learners to MOOCs, allowing
courses to fulfil the promise of "massive" in their titles.

However, the use of these platforms comes with a cost of reducing the pedagogic flexibility allowed to the MOOC authors. Large MOOC platforms are designed to cater to the lowest common denominator with a didactic approach. Other pedagogic approaches are not supported and may indeed be impossible within the constraints of the MOOC platform. If the pedagogic requirements of the MOOC require an approach different from what the MOOC platform provides, going elsewhere may be mandatory.

Cybersecurity: a FutureLearn-based MOOC

The Open University (OU) is the UK's largest university. It offers a range of qualifications from introductory certificates to bachelor's and postgraduate degrees. The OU was founded in 1969 by royal charter with a mission to increase access to higher education. The OU invests heavily in a so-called journey from informal to formal learning by developing learning resources that can be used by casual learners, including television and radio programming, as well as educational material on the Open-Learn platform, iTunes U, and YouTube. These materials are designed to encourage users to begin using small resources before moving on to free self-study courses and MOOCs and eventually to begin formal study towards a university degree.

FutureLearn was founded in 2012 as the first UK-led MOOC platform. It is wholly owned by the Open University but operates as a separate company with its own staff and resources. FutureLearn currently has 40 partners from the UK, Europe, Africa, Asia, and the Middle East. Partners include universities and other learning institutions as well as archival bodies such as museums and national libraries.

Motivation and context for the An Introduction to Cyber Security course

Governments and businesses are gradually becoming aware of the vulnerability of computer networks. Individual awareness of cyber security lags behind that of organisations, with many people simply uninformed of the risks from using a computer. Personal threats include vulnerabilities to bullying and extortion by the release of personal information, as well as the destruction of data by means of malicious software or the improper usage of computers. Individuals of all ages and backgrounds are increasingly vulnerable, and it is necessary to help them acquire the skills to protect themselves from malicious attack as well as accidental damage.

An Introduction to Cyber Security is a free MOOC lasting eight weeks that provides information about cyber security to a non-specialist audience. Learners study key aspects of cyber security and take practical steps to improve their own security. Learners perform security audits to discover the strengths and weaknesses of their own computer systems, develop backup strategies, install security software, and explore the workings of the Internet as well as discussing topical issues with fellow learners and educators.

The cyber security MOOC was funded as a collaboration between the OU's Faculty of Mathematics, Computing and Technology and the United Kingdom government's National Cyber Security Programme managed by the Department of Business, Innovation and Skills (BIS). It forms a significant part of an overall UK government information strategy on cyber security, such as the cyber streetwise campaign (Furnell and Moore, 2014). The material was written, reviewed, and edited by OU staff and reviewed by UK government officials from BIS, the Ministry of Defence, the Cabinet Office, and the GCHQ intelligence agency.

Pedagogy and structure

The course runs four times a year, with every presentation taking eight weeks. Students must create a FutureLearn account to register on the course and access the course materials. Students can join the course up to four weeks after it starts and continue to study after the scheduled course end date (though they will increasingly lack opportunities to discuss the course material with other learners).

The course consists of eight themed weeks of study, with each week intended to take three hours of study by a typical non-expert learner. However, learners are able to adjust their study patterns according to their circumstances, and many take advantage of that flexibility.

The pedagogy of the course is largely dictated by the FutureLearn platform. It is principally a didactic course in which students study course team–prepared material, generally static text and images, supplemented with short animations and video sequences. The material is chunked into small parts within each week; this both increases the flexibility of possible study patterns and allows the materials to be easily studied on a variety of devices (PCs, tablets, and phones). The static material is supplemented with exercises and invitations to discuss the course content in the FutureLearn discussion forums. All material is delivered through the one FutureLearn site.

Learners are expected to complete regular activities, giving them an opportunity to consolidate their learning and apply their knowledge. The activities give learners an opportunity to practice their new skills in a safe, controlled environment (Whitten and Tygar, 1999; Sheng *et al.*, 2006), gain experience of new technologies, and realise how useful they are in real life. Completing each of the activities greatly increases the learner's personal security and collectively, across the cohort, significantly improves the security of the population. The activities include performing a personal cyber security audit, installing various security software packages (antivirus, firewall, and password managers), and using public key cryptography. As learners complete each study task (reading or activity), they mark it complete on the FutureLearn site.

FutureLearn courses are structured to keep learners within the learning environment as much as possible. Links to materials outside the course are minimised and confined to a Links section on each page rather than being embedded within the text. This is a deliberate decision since linking to other sites not only risks learners being directed to a broken site and being unable to continue their studies but also risks learners becoming lost in a maze of pages and unable to return to the course.

Despite this general FutureLearn philosophy of restricting links, learners on the cyber security MOOC are encouraged to supplement the course materials and follow current cyber security developments by regularly reading relevant news and professional websites. The course team suggests a number of accessible sites including the BBC News, *The Guardian*, CNet, and the Open University's own Safe Computing website.

The course was professionally edited by FutureLearn staff to ensure readability and accessibility for a diverse audience of non-specialist novice readers. Technical language was reduced to the minimum required, and a comprehensive glossary of terms was provided for reference.

Assessment

Each week's study has a simple, five-question multiple-choice quiz, automatically marked as the student takes the test. Incorrectly answered questions direct the learner back to the relevant part of the course materials. There is a separate end-of-course assessment, which is another automatically marked multiple choice quiz.

Learners are not required to pass, or even take, any of the assessments. However, if they complete the majority of the learning steps and pass all the tests, learners have the option of buying a certificate of completion. FutureLearn certificates bear the name of the university offering the MOOC (the Open University in this case) but are not considered a university qualification and do not carry any credit towards any university qualification.

Retrospective

The course has now been delivered several times and continues to be presented on the FutureLearn platform. Student numbers for the first four presentations are shown in Table 6.1. In the first year of presentation, more than 73,000 learners signed up to the MOOC, 36,000 completed at least one of the learning activities, and almost 12,000 completed the course. This retention rate of 21% is extremely high for this type of MOOC, where completion rates of 5% to 10% are more common (Adamopoulos *et al.*, 2013). Unfortunately, we do not have more detailed information about partial completions or learner demographics, as that information is retained by FutureLearn for possible future monetisation.

The course materials have been adapted to several other contexts, including presentation in other countries.

By any measure, this MOOC has delivered on its requirements, giving a large number of presumably unskilled members of the public a taste of how to make themselves secure online, and perhaps even taking some simple but effective steps to improve their cyber security at home and work.

Table 6.1 Cyber security MOOC learner numbers

	Run 1		Run 2		Run 3		Run 4		Overall	
		%		%		%		%		%
Joiners	24,330		21,006		14,798		13,175		73,309	
Learners	15,606	64%	12,811	61%	8541	58%	7695	58%	54,815	75%
Active Learners	13,391	86%	10,539	82%	6763	79%	5662	74%	36,355	66%
Returners	8657	55%	6446	50%	3834	45%	3096	40%	22,033	40%
Social Learners	5496	35%	4143	32%	2533	30%	1960	25%	14,132	26%
Full Participants	4280	27%	2873	22%	1766	21%	1311	17%	11,743	21%

The pre-existing MOOC platform allowed the academic staff preparing the MOOC to concentrate on the course content rather than being distracted by evaluating and selecting different components that could be combined to deliver the course. Similarly, the support of editors and artists meant that the learning material was in some cases of higher quality than the academic course team could produce themselves while also saving the academic time.

However, there are a number of problematic aspects to the FutureLearn MOOC production. Most significant is the constraint on pedagogy imposed by the platform. FutureLearn MOOCs are designed to be easily accessible to wide populations; this constrains how sophisticated the learners can be assumed to be and limits the demands that can be imposed on them for learning. This means that MOOC learning is necessarily limited in depth and breadth (courses are encouraged to last no more than eight weeks, with only a few hours of study per week). In addition, the platform only supports a limited number of activities to draw on. Most significant is the restricted functionality of the FutureLearn discussion forums. Different activities have separate and independent forums. Discussions are unthreaded, to ease navigation, but this makes it difficult to follow complex, long discussions. In addition, there are limited features for searching and tagging discussions. These features combine to yield discussions that are good at recording quick responses and interactions but militate against more sophisticated and in-depth discussions.

Another issue is the relationship between FutureLearn and its partners. While wholly owned by the Open University, FutureLearn is a separate commercial entity that has business relationships with many other universities and organisations. FutureLearn is also seeking ways to monetise its student base and learning analytics. This places pressure on FutureLearn to restrict access to the information it has on students and their behaviour, which in turn limits how much MOOC creators can learn about how their MOOCs are received.

Teaching with Tablets: a Blackboard-based MOOC

Much of the content for this MOOC was drawn from the book *Teaching with Tablets* (Caldwell and Bird, 2014) and was intended to allow practising educators to translate current theory into classroom practice. The MOOC was an extension of that idea, with the intent to develop a community of practitioners sharing and learning from each other's practice.

Motivation and context

This MOOC was initiated by the education department in the University of Northampton. It had two main aims. One was to develop a vehicle for disseminating and sharing practice for using tablets (such as iPads) in a variety of educational settings, including schools and higher education institutions (HEIs), and in a variety of disciplines. The other aim was to develop the education department's experience with creating and delivering MOOCs, in particular how such MOOCs can create and sustain communities of practice in educational settings.

The use of mobile devices in education is increasing rapidly and is likely to continue to grow (Ally, 2009). However, new technology poses challenges to educators in that it requires new approaches to teaching and learning (Luckin *et al.*, 2010). To ensure mobile devices enhance learning rather than distract from it, educators need timely guidance on these new approaches. Traditional continual professional development (CPD), based on face-to-face seminars and workshops, can reach only a limited number of educators, whereas an MOOC increases accessibility, giving participants more control over the space, place, and pace of their learning.

Much of the course content was hosted on the University of Northampton's Blackboard server. The same system also handled student registration.

Pedagogy and structure

The MOOC used an innovative, hybridised design that combined features of both x- and cMOOCs in a "structured connectivism" approach that sought to harness the acknowledged power of learning in social settings with the power of a structured design. Online synchronous interactions were combined with asynchronous interactions, and participants were encouraged to collaborate and share examples of their developing practice in an online community space.

With this MOOC, the pedagogy drove the structure and the platform. Existing MOOC platforms, such as the one provided by FutureLearn, were a poor fit to the structured connectivist pedagogy of the Teaching with Tablets MOOC. The intent of the MOOC was to develop a community around the MOOC, where participants were meant to bring much of their own experience to the community and share their experiences with their peers. We deliberately included a range of educational contexts, as we thought they could be useful to all educators. Tablet-based activities and apps intended for young learners could serve as introductory activities for all ages, while more sophisticated activities aimed at older learners could be adapted or serve as inspiration for younger learners.

The MOOC was scheduled to last five weeks, with the course site opening two weeks before the formal course start to allow learners to introduce themselves to the community. We seeded these introductory weeks with simple activities to encourage participants to familiarise themselves with the various apps that would be used often throughout the course.

Each week's study consisted of a reading, two main activities, a number of extension activities, and a Twitter chat. The readings and activities were hosted on the University of Northampton's Blackboard service, and each week's content was only made available from that week onwards. None of the study was compulsory, though participants were encouraged to engage with the reading and at least one of the main activities.

Interaction between learners was important, and most activities in the MOOC required learners to create some artefact using one or more tablet apps and share it with other learners. We created a public community on Google+ for these activities, as it allows learners to create links to online artefacts and comment on their own and those of others. Twitter chats were compiled with Storify and shared online. All these online activities encouraged learners to share their existing expertise and learn from other participants.

Assessment

There was no formal assessment on the MOOC, though learners could buy a certificate of completion. Award of the certificate required that the student could provide evidence of participation in the MOOC, either by showing participation in the Google+ community or other evidence of using tablets in their own learning environment.

Retrospective

The MOOC had 570 students registered, of which 294 accessed the course website and 171 accessed some learning material. The Google+ community had 248 members. The engagement by week shows a reasonably typical drop-off in participation, though 29% of active learners engaged in the fifth week of content (Table 6.2). Figure 6.1 shows how many learners engaged in at least n weeks of

Table 6.2 Engagement by week for Teaching with Tablets MOOC

		of registered	*of engaged*	*of learners*
Registered	570			
Engagers	294	52%	100%	
Learners	171	30%	58%	100%
1 Manipulating media	162	28%	55%	95%
2 Visible learning	86	15%	29%	50%
3 Technology outdoors	68	12%	23%	40%
4 Digital storytelling	57	10%	19%	33%
5 Talk and collaboration	49	9%	17%	29%

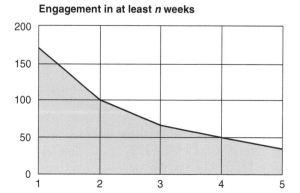

Engagement in at least *n* weeks

Figure 6.1 Numbers engaging in at least *n* weeks of activity in Teaching with Tablets MOOC

the MOOC: of the 171 learners, 50 engaged in at least four weeks and 36 engaged in all five weeks of material. Generally, responses to the MOOC were positive, with many participants saying they found the MOOC useful.

What is not clear from the numbers is the strength of community that developed from the MOOC. All participants drew examples from their own practice, and significant peer learning took place.

Conclusions

The reviews of the two MOOCs should make the differences clear between the two approaches.

The FutureLearn MOOC had the advantage of large reach and support for the academic staff producing the content. However, it had several drawbacks, including a limited choice of pedagogy and constraints on the learning analytics data that was returned to the authors.

In contrast, the Northampton MOOC was much more flexible in its approach, allowing the MOOC to be delivered using a range of tools and platforms to support the most appropriate pedagogy. The details of learners' journeys through the MOOC were more easily captured and analysed, and the staff had a closer relationship with the learners. However, the development of the MOOC required a broader range of skills than with FutureLearn, as the core academic team had to develop all the resources themselves. Finally, the FutureLearn MOOC had a much larger reach than the Northampton one, as FutureLearn was able to publicise the MOOC to its existing base of registered learners. The MOOC had increased reach through the UK government support of the MOOC as part of its cyber security public education efforts.

In conclusion, the correct platform for MOOC development remains open. MOOCs with simple, mainly didactic pedagogies intended for large numbers of learners are best suited to large platforms such as FutureLearn. If the MOOC is intended to serve a more particular audience or requires a more collaborative pedagogy, such large platforms may not be suitable.

Acknowledgements

The MOOCs were developed by a large number of staff at the Open University and the University of Northampton. They include Arosha Bandara, Belinda Green, Anna Cox, Jean Edwards, Jim Atkinson, Kim Calvert, Nicki Wise, Robert Farmer, Wayne Chalmers, and Sway Grantham. We thank them all for their valuable contributions.

References

Ally, M. (2009). *Mobile Learning: Transforming the Delivery of Education and Training.* Edmonton, AB: Athabasca University Press.

Caldwell, H., and Bird, J. (2015). *Teaching with Tablets.* London: Sage.

Conole, G. (2013). MOOCs as Disruptive Technologies: Strategies for Enhancing the Learner Experience and Quality of MOOCs. *Revista de Educación a Distancia*, 39, 1–17.

Furnell, S., and Moore, L. (2014). Security Literacy: The Missing Link in Today's Online Society? *Computer Fraud & Security*, 5, 12–18.

Garrison, D.R., and Kanuka, H. (2004). Blended Learning: Uncovering its Transformative Potential in Higher Education. *The Internet and Higher Education*, 7(2), 95–105.

Hyman, Paul. (2012). In the Year of Disruptive Education. *Communications of the ACM*, 55(12), 20–22.

Khalil, H., and Ebner, M. (2014). MOOCs Completion Rates and Possible Methods to Improve Retention – a Literature Review, in Viteli, J. and Leikomaa, M. (eds.), *Proceedings of World Conference on Educational Multimedia, Hypermedia and Telecommunications 2014* (pp. 1236–1244). Chesapeake, VA: AACE.

Luckin, R., Clark, W., Garnett, F., Whitworth, A., Akass, J., Cook, J., and Robertson, J. (2010). Learner-Generated Contexts: A Framework to Support the Effective. *Web 2.0-Based E-Learning: Applying Social Informatics for Tertiary Teaching*, 70, 70–84.

Nkuyubwatsi, B. (2013). Evaluation of Massive Open Online Courses (MOOCs) from the Learner's Perspective. *Proceedings of the 12th European Conference on e-Learning ECEL-2013*, 2, 340–346, also https://lra.le.ac.uk/handle/2381/28553.

Sharples, M., Adams, A., Ferguson, R., Gaved, M., McAndrew, P., Rienties, B., and Whitelock, D. (2014). *Innovating Pedagogy.* Milton Keynes: The Open University.

Sheng, S., Broderick, L., Koranda, C.A., and Hyland, J.J. (2006). Why Johnny Still Can't Encrypt: Evaluating the Usability of Email Encryption Software, in *Symposium on Usable Privacy and Security* (pp 3–4). New York: ACM.

Stacey, P. (2014). The Pedagogy of MOOCs. *International Journal for Innovation and Quality and in Learning*, (3), 111–115.

Whitten, A., and Tygar, J.D. (1999). Why Johnny Can't Encrypt: A Usability Evaluation of PGP 5.0, in *Proceedings of the 8th USENIX Security Symposium*, Washington DC, [Online]. www.usenix.org/legacy/events/sec99/full_papers/whitten/whitten.ps [Accessed 16th April 2015].

7 Centralised and decentralised virtual lab approaches for information and computer technology education

Peng Li

Introduction

With the rapid development and deployment of information and distributed computing technologies (IDCT), online learning and/or blended learning will play a more and more vital role in academic institutions (Means *et al.*, 2010). It is important to discover and implement effective, efficient, secure, and scalable methods of delivering online, hands-on education. The speed of scientific and technological advances, along with resource constraints, makes it very difficult for schools to maintain technical currency in instructional labs. IDCT makes it possible to share computing capacity and deliver remote hands-on learning experiences effectively and efficiently. IDCT also brings unprecedented collaborative opportunities to bridge the digital divide in education not only between well-equipped schools and less-funded schools but also between developed countries and developing countries. Particularly, in recent years, virtualisation and cloud computing technologies have begun to show the potential to bring transformative changes to the infrastructures of online education in various fields. These technologies provide faculty and students with many opportunities previously unavailable in traditional lab environments.

East Carolina University offers both undergraduate and graduate programmes in information and computer technology (ICT). Our ICT programme prepares students for careers in computer networking, information technology, information security, and technical management. "Learning by doing" is vital for student success. We believe that hands-on, student-centric, active, online learning will be a crucial part of education in this century. Many colleges and universities are facing rising enrolments, while the budgets have not been increased proportionally. In our programme, there are approximately 200 students. Enrolment has gone up in the past few years, especially in the distance education (DE) section. How to use the limited resources to deliver quality hands-on education effectively and efficiently becomes a real challenge.

Our exploration progressed from the traditional, centralised, physical lab environment to a decentralised virtual lab environment and then back to the centralised virtual lab supported by cloud computing. We have been exploring multiple ways of delivering virtual labs, including decentralised approaches and different centralised approaches, since 2006. Student survey feedback has been positive and encouraging. In a typical decentralised approach (Li, 2009), students perform lab experiments using a hosted hypervisor such as VMware Workstation or Oracle xVM VirtualBox on their personal computers. In centralised approaches (Li, Jones and Augustus, 2011), the virtual lab experiments are hosted on one or multiple servers on campus, utilising different private cloud systems within the university. We have developed multiple virtual labs and virtual environments, which are unique, innovative, and effective for active student learning.

Virtualisation and cloud computing

Virtualisation technology, also known as virtual machine (VM) technology, allows multiple virtual machines to run concurrently and independently on a single physical computer. A typical virtual environment consists of one or more virtual machines, often contained inside a sandbox: a tightly-controlled environment with highly restricted network access. Virtualisation and cloud computing technologies have been extensively adopted in industry because they make it possible to provision computing resources and services more efficiently and effectively. In education, these technologies provide faculty and students with many opportunities previously unavailable in traditional lab environments.

Virtualisation technology was first developed at IBM and used on their mainframes in the 1960s (Creasey, 1981). However, there was not much development on x86 platforms until the late 1990s (Rosenblum, 2004). In 1999, VMware released the first x86 virtualisation product. The virtualisation competition gradually increased, and different virtualisation projects were initiated in the early 2000s.

Virtual machines were used in ICT education as early as 2003 (Nakagawa *et al.*, 2003; Stockman, 2003). Virtualisation was intended to help the reduction of hardware and maintenance costs, improve the availability of lab resources, and expedite the deployment of new technologies. Most early virtualisation implementations involved a centralised approach: virtual machines were hosted on central servers or lab computers on campus; students connect to virtual environments remotely to perform labs.

The centralised virtual lab (Stockman, Nyland and Weed, 2005; Gaspar *et al.*, 2008; Stackpole, 2008) as the replacement of the physical lab has been found to be successful in reducing costs and in improving availability and performance. Some early studies indicated that students using virtual machines to do laboratory work performed as well as students using physical machines in a traditional computer lab (Lawson and Stackpole, 2006).

Aims of the project

In a traditional centralised physical remote lab, physical machines and equipment are typically housed in a machine room on campus. The computers and instruments are connected through hubs or switches to a central server. Students connect to the central server remotely to conduct lab exercises.

The centralised remote physical lab approach makes it possible for distance-education students to perform hands-on labs online. It is still irreplaceable in some fields because not all equipment can be virtualised. However, the traditional approach has disadvantages, including but not limited to (1) the lab equipment, such as physical computers, switches, cables, routers, among other things, is expensive; (2) the cost of maintenance can be high; (3) remote access to these labs can be slow and unstable, and (4) students often have to share the equipment with others. Those who are unable to complete their exercises within the designated time slots may not be able to continue their work.

Virtualisation technology enables multiple isolated virtual machines with guest operating systems to run simultaneously and independently on a single physical machine. The technology significantly reduces or completely eliminates the necessity to have multiple computers that host diverse operating systems or networking services typically deployed in physical labs. Advances in virtualisation and distributed computing have facilitated the development of cloud computing services. Cloud computing is a general model for delivering information technology (IT) services and applications, on demand, over a private or public network. Virtualisation and cloud computing technologies can help the reduction of hardware and maintenance costs, improve the availability of lab resources, and expedite the deployment of new technologies.

The concept of "virtual environment" in this chapter is different from that of "virtual environment" in a virtual-reality simulator such as Second Life or Open-Sim in that virtualisation is employed to run and deliver real operating systems, computer applications, or network services. Generally speaking, virtualisation is neither simulation nor emulation. A typical virtual environment consists of one or more virtual machines, often contained inside a sandbox – a tightly controlled environment with highly restricted network access. Real operating systems such as Windows, Linux, and FreeBSD can run seamlessly on a virtual machine. A virtual machine can host real applications or services such as Apache Web Server and Oracle Database Server.

Virtualisation technology was first developed at IBM in the 1960s to offer concurrent timesharing of their mainframes (Creasy, 1981). Virtual machine monitor (VMM) was introduced as a software abstraction layer to manage the mainframe hardware and to support virtual machines. Each virtual machine (VM) runs like a physical machine. Users can install operating systems and execute applications in the virtual machines in the same manner as they would on physical machines.

It was not until the late 1990s that virtualisation technology breakthrough occurred on the Intel x86 platforms, which allowed users to run multiple virtual machines (including operating systems and applications) simultaneously on a single personal computer (Rosenblum, 2004). Most early virtualisation implementations involved a centralised approach (Stockman, Nyland and Weed, 2005; Gaspar *et al.*, 2008). Virtualisation programs such as VMware were costly at that time. The computers running virtual machines required a lot of RAM, which was expensive. As a result, virtualisation software was not very popular for personal use. Virtual machines were usually hosted on central servers or lab computers on campus. Students either sat in a computer room to use them or connected remotely to an on-campus lab to use them.

Online education has witnessed tremendous growth in the past decade. Advances in technologies made it possible to deliver not only lectures but also hands-on labs remotely. Traditional online courses were heavily weighted toward lecture and paper assessment, justified because of lab hardware cost, lab footprint, lab availability, and inability to objectively measure student lab performances. While cost remains a factor in any course with a lab component, implementing new technologies such as virtualisation and cloud computing has mitigated and in some cases completely removed ancillary lab issues. Consequently, online cloud-based lab experiments are being developed for use by both distance education (DE) and face-to-face (F2F) students.

To that end, we have been working on projects that integrate virtualisation and cloud technologies to provide concrete online learning experiences to students. This will, among other things, allow us to

1   set up private cloud infrastructures using Virtual Computing Lab (VCL), VMware Lab Manager (VLM), and ProxMox VE to host virtual labs;
2   create custom virtual environments and hands-on information technology lab modules which can be hosted on cloud/grid computing networks in a central-ised manner and on students' personal computers in a decentralised manner; and
3   evaluate different centralised and decentralised virtual lab approaches for educational purposes.

Virtualisation methods

Virtualisation technology allows a single physical computer to run multiple iso-lated virtual machines concurrently. Virtualisation is implemented in different ways. Three types of virtualisation methods are used in this project: type I vir-tualisation, type II virtualisation, and container-based virtualisation. Different hypervisors are used in type I virtualisation and type II virtualisation. A hyper-visor, otherwise known as virtual machine monitor (VMM), enables multiple isolated guest operating systems to run simultaneously on a host machine. There are two types of hypervisors: type I (native or bare metal) hypervisors and type II (hosted) hypervisors. Native (bare metal) hypervisors run directly on the physical machine to host the guest virtual machines. Notable native hypervisors include but are not limited to Xen, KVM (Kernel-based Virtual Machine), VMware ESX/ESXi, and Microsoft Hyper-V. Hosted hypervisors run as applications on top of a commodity operating system such as Windows, Linux, or UNIX. Notable hosted hypervisors include Oracle VM VirtualBox (formally known as Sun xVM VirtualBox), VMware Workstation, VMware Server, VMware Fusion, Parallels Workstation, QEMU, and Microsoft Virtual PC, among others. Both types of hypervisors usually support diversified, iso-lated guest operating systems. For example, VMware ESXi (a native hypervi-sor) and Oracle VM VirtualBox (a hosted hypervisor) can host virtual machines running different Windows, Linux, and BSD guest operating systems. Native hypervisors generally have higher efficiency (performance and scalability) than hosted hypervisors and are used in building cloud computing infrastructures. Unlike hosted hypervisors, native hypervisors are installed directly on physical machines without operating systems, and they normally cannot be installed on top of unmodified operating systems such as Windows XP. Hosted hypervisors such as VMware Workstation are installed under existing Windows/Linux operating systems, which is convenient for certain scenarios, for example home use or development/testing.

The container-based virtualisation, also called operating system–level virtualisation, uses a shared operating system (OS) image. The containers (also known as virtual private servers, or VPS) are isolated OS instances. A container behaves like a real server but runs a virtualised OS instead of a real OS. Container-based virtualisation, featuring very low overhead, performs and scales better than hypervisor-based virtualisation. However, container-based virtualisation does not support diversified guest operating systems. If the host OS is Linux, the container OS must be of the same type. The examples of container-based virtualisation solutions include but are not limited to Solaris Zones, FreeBSD Jails, Linux VServer, and OpenVZ (the basis of Parallels Virtuozzo Containers).

With increasing enrolment in distance education, the high demand for remote labs significantly impacted our already limited resources. In 2006, we began incorporating virtualisation technology in IT labs in order to find cost-effective ways of delivering remote hands-on exercises. In the first three years, a decentralised virtual lab approach was implemented: the students installed prebuilt virtual machines on their personal computers to complete hands-on labs. Various virtualisation software packages were used in successive years and were focused on hosted hypervisors, which depended on a host operating system to run the virtualisation software, beginning with VMware Player and VMware Server, followed by VMware Workstation and Sun xVM VirtualBox (now renamed Oracle VM VirtualBox). The student feedback was positive, and the student performance was consistent.

Then we started experimenting with centralised approaches to support virtual labs which require high-end hardware/software. The first cloud infrastructure we used was Virtual Computing Lab (VCL), an open source, on-demand, remote access system that dynamically provisions computing resources to end users (Averitt, 2007). Then we deployed VMware Lab Manager and ProxMox Virtual Environment (VE) Server in different courses.

Decentralised virtual lab approach

In a decentralised virtual lab approach (Li, 2009), the labs are hosted not on central servers on campus but on students' personal computers. Typically, preconfigured virtual machines and other lab materials are prepared by the instructors and distributed to the students on optical media such as CD or DVD or through Internet downloads. The students install the virtual machines on their personal computers and complete hands-on exercises using the virtual machines. The students have the option but do not have to connect to a central server on the university campus.

In the first trial with the virtual lab in 2006, only one pre-built Fedora Core 5 Linux VM was used due to the concern that some students' computers might not be powerful enough to handle a second VM. The virtual machine was created by the instructor and made available on the class ftp server for download. Students used freeware VMware Player or VMware Server to run the pre-built virtual machine and perform labs. Next year, two pre-built virtual machines were provided to the students. VM1, running Debian Linux 3.1, acted as the client or the attacker, from which the students launched scans and attacks. The intrusion detection system (IDS) sensor was installed on VM1. VM2, running CentOS Linux 5.0, acted as the server or the target, on which the students set up the defence. Students used VMware Workstation to run the virtual machines. Some projects were redesigned accordingly to accommodate the two VMs. Efforts were made to make the virtual lab more portable. The Linux VMs were customised to reduce their sizes. In the third year (2008), VMware Workstation was replaced by Sun xVM VirtualBox as the hosted hypervisor to run the virtual machines. Compared to fall 2007, the portability of the virtual lab was further improved. The virtual lab now could run on different platforms such as Windows XP, Windows Vista, and Linux. The licence of VirtualBox was less restrictive, so potentially more people (e.g. in community outreach programmes) can benefit from running the virtual lab. More customisation was implemented on the VMs, and we were able to fit all lab files on a 700 MB CD-R instead of a DVD-R.

The decentralised virtual lab approach has the following advantages: (1) it is portable because it allows students to conduct hands-on lab exercises at any place, at any time, and at their own pace, so they no longer need to share lab equipment with other students; (2) the virtual lab is expandable. As long as the host machine has sufficient memory and storage, virtual machines can be duplicated and added easily; (3) because the virtual machines run locally, they are much more responsive than the physical/virtual machines accessed remotely; (4) the hardware cost of setting up the virtual lab is low since the labs are conducted on students' personal computers; and (5) the maintenance is relatively easy since corrupted virtual machines can be restored or replaced within a very short period of time.

The decentralised virtual lab, however, has its limitations: (1) not all lab equipment can be virtualised; (2) a personal computer may not be powerful enough to support multiple virtual machines; and (3) license and security restrictions may prevent some labs from being performed on personal computers. To that end, the decentralised virtual lab approach is complementary to the centralised lab approaches but will not replace them in the foreseeable future.

Centralised virtual lab approaches

The decentralised virtual lab approach was cost effective; however, it did not scale well for labs requiring more than three virtual machines. For example, a virtualised datacenter lab environment used in one course was composed of five virtual machines, using approximately 8 GB RAM and 44 GB hard drive space. Not all students' personal computers were powerful enough to run so many virtual machines efficiently. From 2008, we started experimenting with new centralised virtual lab systems, specifically Virtual Computing Lab (VCL), which was used in several projects that year. The next year, VMware (vCenter) Lab Manager (VLM), another virtual lab automation package, was made available in selected courses.

Virtual Computing Lab

Virtual Computing Lab (VCL) is an open source, on-demand, remote-access system that dynamically provisions computing resources to end users. VCL is a cost-effective implementation of cloud computing (Dreher, 2010; 236). The system was initially developed at North Carolina State University and became an Apache incubator project in 2008. VCL is different from the traditional centralised remote lab systems in that the virtual environments (images) are stored in the image library and are loaded, on demand, on physical or virtual machines for student and faculty use. A typical VCL infrastructure is composed of a web front-end including the scheduler and the database, a management node, an application image (virtual environment) library, and a server center. The user makes a reservation for a virtual environment through the VCL web site. The scheduler will check whether the virtual environment (VE) is already available to the user. If not, the scheduler will instruct the management node to retrieve the VE image from the application image library, load it on a virtual machine or a physical server, and make it available to the end user. The user may access the virtual environment using different protocols, including RDP, SSH, and VNC. After the reservation ends, the virtual machine or physical server will be released and made available to host virtual environments for other users. A VCL server can be installed on a Linux virtual machine or physical computer and can provision (allocate) not only virtual machines but also physical computers. VCL can be deployed with low-cost commodity hardware but is particularly suitable for labs requiring high-end hardware and/or software. VCL is also useful for collaborative learning. Students in team projects can work together on remote VCL computers through VNC or Remote Desktop.

Among the courses in which Virtual Computing Lab was used, the labs in the course *Virtualization Technologies* were most resource demanding. If physical hardware is used, the minimum server requirement for a single student would be: two 64-bit processors, 4GB or more RAM, three Ethernet NICs (100MB or 1 GB), and 120 GB local storage accessible through the SCSI controller. For a typical section of 24 students, a minimal of 12 dedicated student servers were required, in addition to an infrastructure server, a data storage server, and gigabit Ethernet switches, cabling, and, optionally 24 student workstations. It would be very expensive to set up and maintain a physical lab for this class.

Instead, we created a custom virtual datacentre environment consisting of five virtual machines hosting two VMware ESX4 servers, a vCenter VM, and an Openfiler iSCSI/NFS VM, and a client workstation, respectively. The virtual datacenter was created as a VCL image for students to reserve. More details about this virtual environment were published elsewhere (Li, 2010a). Without the virtual machine solution, it is unlikely students would have the opportunity to do the hands-on labs due to the high cost of physical labs.

VMware Lab Manager

VMware vCenter Lab Manager (VLM) is a commercial product by VMware Inc. Educational institutions may obtain free licenses through the VMware Academic Program. The backend of VLM is a group of VMware ESX/ESXi servers, controlled by a VMware Virtual Center or vCenter. VLM can only be used to provision VMware virtual machines and configurations. It provides user-friendly interfaces for instructors/administrators to manage users, roles, workspaces, and configurations and allows users to access the virtual machine consoles remotely. VLM offers some powerful features not yet available on other VLA systems such as fencing. VLM offers persistent storage so students can save their work and come back later to continue.

In 2010, VMware vCenter Lab Manager (VLM) was made available to students in ICTN 4200/01 Intrusion Detection Technologies. Each student was assigned a configuration with three virtual machines: the first VM as the defender, the second as the attacker, and the third as the client workstation. The configuration could be expanded, and more VMs could be added easily. The students were assigned 12 required labs and a number of optional, bonus projects. The students accessed the VLM server using a web browser (Internet Explorer or Mozilla Firefox) and then logged in to the client workstation (a Windows XP VM). From there, they could access the other two Linux virtual machines through SSH. They could also access the consoles of the virtual machines. After the students finished certain tasks, the changes they made were automatically saved, enabling students to come back later to continue the lab. With VLM, the administrator/instructor was able to monitor the lab operations of every student and provide assistance via a web browser remotely.

Container-based centralised virtual lab

Traditionally, students studying web services programming and management were assigned accounts on a shared web server with very restricted privileges due to security concerns and limited resources. Consequently, the students' experiences were restricted only to those that could be safely supported by the server due to the limitations imposed by the shared environment. OS–level virtualisation, also called container-based virtualisation, allows a single well-equipped physical computer to run many instances of virtual private servers. The OS–level virtualisation is more efficient and scalable than some other virtualisation options. The technology makes it possible for every student to have her/his own virtual private server, on which she/he can carry out programming/administration tasks with full privileges. It is important for the ICT students to understand how to use virtualisation to deploy IT applications and services.

Different lab automation systems, including ProxMox, among others, were used to host the OpenVZ servers for OS–level virtualisation. The newly created lab modules were used selectively in ICTN 2732 Scripting for Information Technology and ICTN 3900/01 Web Services Management. Both the course content and corresponding lab manuals were developed/updated to reflect the new learning environment. The deployment of OS-level virtualisation provides a cost-effective way to develop students' hands-on skills and can be adopted for use in other information and computer technology courses.

Citrix XenDesktop

Citrix XenDesktop is a virtualisation solution that delivers on-demand Windows desktops to end users. XenDesktop is not as flexible as Virtual Computing Lab in the variety of custom virtual environments but is useful for remote labs which require students to use high-end software packages such as Adobe Photoshop.

Student assessment within a virtualised environment

It is challenging for educators to assist, monitor, and objectively evaluate students in remote lab environments. The asynchronous assessment approach is commonly used, in which students submit completed answer sheets, lab reports, scripts, and/ or other supporting materials for the instructor or the peer to review. The synchronous assessment approach allows immediate feedback. After a lab task is completed, the student can initiate the grading process by clicking a hyperlink on a website. A data collection script will be forwarded to her/his virtual machine. Lab results are gathered on the virtual machine and returned to the central server for automated or manual assessment and feedback.

BroadReach is a locally developed solution by Mr Lee Toderick which facilitates both asynchronous and synchronous assessments (Toderick, 2007). It manages student access through an IPSec virtual private network (VPN) between distributed nodes and a central server. In a typical lab, a student logs in to a node (which can be a virtual machine on her/his personal computer, a physical machine on campus, or a computing node in the cloud), run the BroadReach VPN setup script to establish the connection with the VPN gateway server, and then perform hands-on exercises on the node. Through the VPN, a teacher can access the remote node, observe student activities, and provide real-time assistance when needed. The teacher can run scripts to collect lab results from the remote node and send the results back to the central server for grading. Students can download the grading scripts through the VPN connection and perform self-assessments.

We are planning to extend the system to provide real-time instructor assistance when needed through the use of collaboration tools; deliver hands-on, performance-based exams that demonstrate student competency in the subject matter; enable students to submit self-assessment deliverables to a central server for immediate grading; and collect assessment statistics data and generate reports on students' critical learning rubrics.

Outcomes of the study

Since 2006, we have been exploring different virtual lab approaches in the Information and Computing Program at East Carolina University. The goal is to provide an online, scalable virtual lab framework so both F2F and DE students can study at any place, at any time and at their own pace.

What we did *not* do in this research is to set up and to compare two control groups of students. One uses physical labs only and the other uses virtual labs only. As mentioned previously, many physical labs are so resource demanding that it is cost prohibitive to implement them in most academic institutions. Virtualisation makes it possible for us to deploy these labs in a centralised/decentralised virtual environment. Virtualisation provides our students with opportunities which are otherwise not available in the traditional, physical labs. Student assessment results show that learning objectives are achieved. A rough comparison shows no significant differences in achievements between students using decentralised virtual lab approaches and those using centralised virtual lab approaches (Li, 2009; Li, Jones and Augustus, 2011).

The direct outcomes of the virtual lab project include:

Outcome #1: More than 25 virtual labs have been developed and deployed in six courses, including ICTN 2732, ICTN 3900/01, ICTN 4200/01, ICTN 4700/01, ICTN 4750/01, and ICTN 4800/01. Some labs are portable between centralised and decentralised environments. Other labs require high-end hardware/software and are typically hosted in the centralised private cloud environments. Many lab modules are portable between courses and can be used by faculty and students in different courses and in different schools.

Outcome #2: Student surveys were conducted in several courses and showed positive feedback. The following tables provide snapshots of the feedback. More details about the results are published in other papers (Li, 2009; Li, Jones and Augustus, 2011).

An anonymous, discretionary survey was conducted online at the end of the spring 2011 semester in ICTN 2732. Seven out of 19 students responded. An anonymous, optional survey was administered online at the end of the fall 2010 semester in ICTN 3900/01. Fifteen out of 30 students in three sections responded. Tables 7.1 and 7.2 summarise the responses for centralised virtual labs. Some rows did not add up to 100% because a student left some questions unanswered.

An anonymous, optional survey was conducted online at the end of the fall 2008 semester in ICTN 4200/01. Sixteen out of 20 on-campus students and 30 out

Table 7.1 Centralised virtual labs in ICTN 2732

ICTN 2732 Section 001 (Centralised virtual labs)	Strongly Agree	Agree	Neutral	Disagree	Strongly Disagree
The hands-on homework projects facilitated my understanding of the course topics.	85.7%	14.3%	0%	0%	0%
The instructions in hands-on homework projects were easy to understand.	57.1%	42.3%	0%	0%	0%
The Virtual Private Server (VPS) was easy to use.	71.4%	28.6%	0%	0%	0%
I had access to the Virtual Private Server (VPS) when I needed to use it.	85.7%	14.3%	0%	0%	0%
The share class web server was easy to use.	85.7%	14.3%	0%	0%	0%
I had access to the share class web server when I needed to use it.	85.7%	14.3%	0%	0%	0%

Table 7.2 Centralised virtual labs in ICTN 3900/01

ICTN 3900/01 Sections 001/301/601 (Centralised virtual labs)	Strongly Agree	Agree	Neutral	Disagree	Strongly Disagree
The labs facilitated my understanding of the course topics.	66.7%	33.3%	0%	0%	0%
The lab instructions were easy to understand.	60%	33.3%	0%	0%	0%
The virtual lab environment (proxmox/win2003/win2008/winxp) was easy to use.	66.7%	33.3%	0%	0%	0%
The hands-on exercises in the virtual lab environment were as effective as those in physical computer labs.	60%	26.7%	13.3%	0%	0%
VMware Lab Manager was easy to use.	73.3%	26.6%	0%	0%	0%
I had access to VMware Lab Manager when I needed to use it.	73.3%	20%	6.7%	0%	0%
Virtual Computing Lab (VCL) was easy to use.	40%	33.3%	13.3%	6.7%	0%
I had access to Virtual Computing Lab when I needed to use it.	40%	40%	13.3%	0%	0%

Table 7.3 Decentralised virtual labs in ICTN 4200/01

(Decentralised virtual labs)	Strongly Agree	Agree	Neutral	Disagree	Strongly Disagree
The labs helped me understand the course topics.	50%	43%	4%	0%	0%
The lab projects were well organised.	67%	26%	2%	2%	0%
The lab instructions were straightforward.	63%	26%	4%	4%	0%
The hands-on exercises in the VirtualBox virtual lab were as effective as the ones in a physical computer lab.	72%	26%	0%	0%	0%
The pre-configured virtual machines were easy to use.	74%	20%	2%	2%	0%
The demo Flash videos were helpful in learning.	30%	30%	34%	2%	0%
I was able to learn new skills and technologies through the virtual lab.	48%	39%	9%	0%	2%
I was able to study at my own pace using VirtualBox.	54%	41%	0%	2%	0%
The virtual lab was time-consuming.	22%	24%	35%	17%	0%

of 41 distance-education students responded. Table 7.3 summarises the responses for decentralised virtual labs. Some rows did not add up to 100% because a student left some questions unanswered.

Outcome #3: Supported by several grants including HP Catalyst Initiative, we have been able to install and use in-house private cloud systems to test and conduct research into the effectiveness of different virtual lab approaches. Cloud computing can be implemented in many different forms and complexities. Cloud systems such as VMware Lab Manager and Virtual Computing Lab (VCL) provide innovative, complementary, but inherently different ways to deliver hands-on exercises in a centralised manner. For example, Lab Manager virtual machine configuration changes are persistent, while VCL virtual machines lose configuration changes at the end of a reservation period. The cloud approach is particularly useful in remote labs that require high-end hardware/software, support a large number of students, and provide constantly evolving contents. An independent VMware vCenter Lab Manager system was installed using HP ProLiant servers. The network-attached storage was managed by HP Virtual Storage Appliance, also running on HP ProLiant servers. The ProxMox VE server was hosted on a separate HP ProLiant server.

Outcome #4: Our collaboration with HP Catalyst partner institutions on virtual labs is still in its infancy but ongoing. The initial experiences, however, indicate that administrative involvement/awareness and strong IT support staff are essential to successful collaborations. In addition to the cloud infrastructures, developing relevant labs contents is also crucial. We plan to continue designing, testing, and deploying new virtual labs and custom virtual environments which are portable between different courses, different platforms, and different institutions through the assistance of public and private grant funding.

There are potential pitfalls to adopting cloud solutions. For example, setting up or maintaining a cloud system is not an easy task. A capable technical support team is important for successful deployment of virtual labs, especially the ones utilising centralised cloud computing systems. This team needs to install, maintain, and troubleshoot the equipment, the software, and the custom virtual environments. We found university resource requirements for deploying a decentralised virtual lab to be less demanding. The users (students), however, need to have personal computers powerful enough to run one, sometimes several, virtual machines. Also, the instructor needs to have basic knowledge and skills in setting up and managing virtual machines. In addition to the free, open-source Virtual Computing Lab and ProxMox Virtual Environment, we use VMware vCenter Lab Manager, which is a commercial, proprietary software program. Other educational institutions are advised to join VMware Academic Program (www.vmware.com/partners/academic/program-overview.html) to receive free licenses for VMware vSphere and Lab Manager. We believe that any institutions adopting virtualisation will find it useful. It is likely that the new technologies will provide new learning experiences which would be unavailable otherwise.

There are still many unexplored or underexplored areas in the research of integrating virtualisation and cloud computing technologies in online hands-on education. For instance, how can we conduct effective assessment in a cloud computing environment? What are the cost factors of different cloud computing systems designed for educational purposes? What impact does this environment have on STEM education?

We have two different groups of students: face-to-face students and distance education students. However, not all courses are available for the two groups at the same time. We will consider studying the potential different effects in the future.

Conclusions

Decentralised and centralised virtual lab approaches have been investigated and implemented in information and computer technology courses using different virtualisation and cloud computing techniques. Hosted hypervisors are suitable for decentralised virtual labs performed on students' personal computers. The decentralised virtual lab approach, with no need to reserve or share equipment, provides maximum availability. Centralised virtual labs can be deployed using native hypervisor-based technology and/or container-based technology. The centralised virtual lab approach, supporting high-end software and/or hardware, allows fast deployment of new labs and facilitates team cooperation. The decentralised and centralised virtual lab approaches are not only viable but also cost-effective alternatives to the traditional centralised physical lab approach. However, they can also be complementary to each other.

Centralised virtual lab solutions allow users to access remote labs more easily, without the need to install virtual machines on their local computers. The centralised systems can support virtual labs requiring multiple virtual machines and other resources. It is easier for the instructor to monitor the lab activities and for students to seek help or to collaborate in a centralised environment. Furthermore, the collection and analysis of usage data to aid effective resource allocation is made easier within a centralised setting. Our experiences with Virtual Computing Lab (VCL) and VMware Lab Manager (VLM) show they are both flexible in supporting hands-on IT labs in a variety of areas. VLM is easier to manage and more user friendly and efficient because it only provisions virtual machines. VCL is open source and has the ability to provision both virtual machines and physical computers. The decision on virtual lab approach or virtual lab automation system to use depends on the needs of the course and the available resources.

Cloud computing can be implemented in many different forms. Centralised virtual lab systems such as Virtual Computing Lab and VMware Lab Manager provide cost-effective ways of delivering hands-on learning experiences in a centralised manner. The centralised approach is especially useful in online labs which require high-end resources, support a large number of users, and provide rapidly changing contents. It helps create a digital learning environment for students to study from any place and at any time.

References

Averitt, S., Bugaev, M., Peeler, A., Shaffer, H., Sills, E., Stein, S., Thompson, J., and Vouk, M. (2007). Virtual Computing Laboratory (VCL). *Proceedings of the International Conference on the Virtual Computing Initiative*, IBM, 1–6.

Creasy, R.J. (1981). The Origin of the VM/370 Time-Sharing System. *IBM Journal of Research and Development*, 25(5), 483–490.

Dreher, P., Vouk, M.A, Sills, E., and Averitt, S. (2010). Evidence for a Cost Effective Cloud Computing Implementation Based Upon the NC State Virtual Computing Laboratory Model, in Gentzsch, W. Grandinetti, L. and Joubert, G. (eds.), *High Speed and Large Scale Scientific Computing* (pp. 236–250). Amsterdam: IOS Press.

Gaspar, A., Langevin, S., Armitage, W., Sekar, R., and Daniels, T. (2008). The Role of Virtualisation in Computing Education. *SIGCSE Bull*, 40(1), 131–132.

Lawson, E.A., and Stackpole, W. (2006). Does a Virtual Networking Laboratory Result in Similar Student Achievement and Satisfaction? *Proceedings of the 7th Conference on Information Technology Education*, ACM, 105–114.

Li, P. (2009). Exploring Virtual Environments in a Decentralised Lab. *Research in Information Technology*, 6 (1), 4–10.

Li, P., Jones, J., and Augustus, K. (2011). Incorporating Virtual Lab Automation Systems in IT Education. *Proceedings of 2011 ASEE Annual Conference and Exposition*, ASEE (pp. 22.856.1–22.856.13).

Li, P., and Toderick, L. (2010b). Cloud in Cloud: Approaches and Implementations. *Proceedings of 2010 ACM SIGITE Conference*, ACM, 105–110.

Li, P., Toderick, L., and Noles, J. (2010a). Provisioning Virtualised Datacenters through Virtual Computing Lab. *Proceedings of 2010 ASEE/IEEE Frontiers in Education Conference* (pp. T3C-1–T3C-6), Arlington, Virginia: IEEE.

Means, B., Toyama, Y., Murphy, R., Bakia, M., and Jones, J. (2010). Evaluation of Evidence-Based Practices in Online Learning: A Meta-Analysis and Review of Online Learning Studies. *Center for Technology in Learning*, U.S. Department of Education. http://eric.ed.gov/?id=eD505824.

Nakagawa, Y., Suda, H., Ukigai, M., and Miida, Y. (2003). An Innovative Hands-on Laboratory for Teaching a Networking Course. *Proceedings of 33rd Annual Frontiers in Education Conference*, IEEE, 1, T2C-14–20.

Rosenblum, M. (2004). The Reincarnation of Virtual Machines. *Queue*, 2(5), 34–40.

Stackpole, B. (2008). The evolution of a virtualised laboratory environment. *Proceedings of the 2008 ACM SIGITE Conference on Information Technology Education*, ACM, 243–248.

Stockman, M. (2003). Creating remotely accessible "virtual networks" on a single PC to teach computer networking and operating systems. *Proceedings of the 4th Conference on Information Technology Curriculum*, ACM, 67–71.

Stockman, M., Nyland, J., and WEED, W. (2005). Centrally-Stored and Delivered Virtual Machines in the Networking/System Administration Lab. *SIGITE Newsl*, 2(2), 4–6.

Toderick, L., and Lunsford, P. (2007). Using VPN technology to remove physical barriers in Linux lab experiments. *Proceedings of the 8th ACM SIGITE Conference on Information Technology Education*, ACM, 113–118.

8   Learning to create a better built environment

Giovanni C. Migliaccio, Ken-Yu Lin and Carrie Dossick

Introduction

The College of Built Environments (CBE) at the University of Washington (UW) believes that university curricula should be designed to promote critical thinking and problem solving that target real world issues. Built environment subjects at UW include construction management (CM), architecture (Arch), real estate (RE), urban design and planning (UDP), community, environment, and planning (CEP), and landscape architecture (LA). Students in these programs are required after graduation to face some of the most imminent challenges, including urbanisation, globalisation, and sustainability.

As such, faculty members with the University of Washington's Department of Construction Management are working jointly with other CBE colleagues and administrators to experiment with new curricula and embed experiential and contextual learning concepts into their courses to bridge the gap between what students learn in class and how they can effectively apply that knowledge in settings outside of school. This objective is traditionally easier to implement in some topics than in others. For example, learning about construction materials and methods can be enhanced by performing experiential activities in a physical lab. Similarly, learning about construction scheduling and estimating can be achieved through the use of cases contextualised on real projects and simulation of bidding and scheduling activities in a computer lab with latest and up-to-date technology.

However, offering in-class or in-lab experiential learning opportunities is more difficult on other topics, such as *geographically dispersed teamwork* and *construction safety*. Still, these learning activities are needed to prepare students in built environment careers. As design and construction become more global, professionals entering the workplace need to be prepared to work across geographical distances with industry partners. Traditional classroom settings where students and instructors are co-located provide little to no opportunity to explore tools and work processes for geographically distributed work. Consequently, students in design-build courses would greatly benefit from being exposed to teamworking with students in other institutions as members of geographically dispersed teams. Furthermore, students in construction safety courses would need exposure to real hazardous situations to assess their readiness in recognizing and addressing construction site hazards. However, liabilities associated with this "experience" would make challenging to design and implement a learning event, such as a construction site walk-through.

Financial support from the HP Catalyst and HP Leadership Fund have offered the UW's Construction Management department an opportunity to design, experiment, and implement technology solutions to address globally distributed work and construction safety curriculum and allowed them to deliver experiential and contextual learning events in a wide range of courses. The project set up under this initiative entails a number of activities designed to answer several questions, including:

1   Do cloud computing and technology facilitate the learning of the skill and strategy needed for global collaboration in regards to the built environment disciplines like urbanisation, globalisation, and sustainability?
2   Does access to technology facilitate experiential and contextual learning?

Since the HP technology was received in October 2011 and the HP Leadership Fund was awarded in 2012, we initiated several activities. Examples of completed activities facing these questions are presented in the following sections.

Do cloud computing and technology facilitate the learning of the skill and strategy needed for global collaboration?

ECL+VI Construction Catalyst

With support by the HP Leadership Fund, the College of Built Environments at UW (UW-CBE), the Masinde Muliro University of Science and Technology (MMUST), industry partners, selected K–12 schools in the Seattle area, and selected rural schools in Kenya carried out several initiatives under a programmatic umbrella named the ECL+VI Construction Catalyst. Relying on experiential and contextual learning (ECL), this collaborative effort planned to use a construction project-based theme to deliver activities for innovating STEM+ learning and teaching. The programme "grafted" the vertical integration (VI) approach used by MMUST into the ECL approach used by the Department of Construction Management at UW (UW-CM). While facilitating a learning-in-context of STEM+ topics, the proposed ECL+VI approach was also intended to help K–12 students to decide a career-path at the university level matching their tendency toward contextual or abstract learning. The vision of the ECL+VI programme is shown in Figure 8.1.

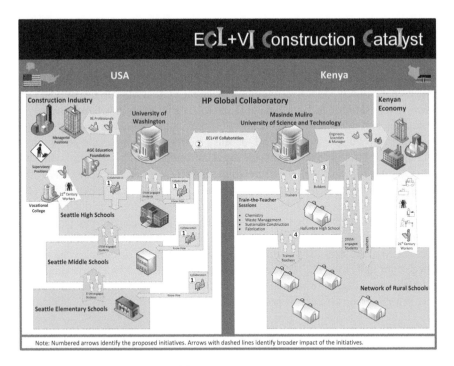

Figure 8.1 ECL+VI programme vision

The UW-CM programme has an alumni community of roughly 2,000 professionals, which includes executives in many major companies. ECL has been a pervasive methodology in the UW-CM programme. This exposure to staged experiential opportunities is one of the factors attributed to making students of this programme highly successful once they graduate and as they assume junior managerial and project engineering positions in the industry.

Traditionally, STEM+ initiatives have used the VI approach with a focus on engineering or the sciences. As a result, various engineering- and science-oriented events, activities, and materials are available to students. This approach may result in STEM-engaged students considering only these two educational paths in STEM+ disciplines. Once they are in college, a large number of these students find that the engineering and science curricula are often too abstract and decontextualised. Many of these students lack engagement, and some encounter academic problems. Many of the current and former students in the UW-CM programme are former engineering and science students who found the ECL environment used in the UW-CM programme to be more engaging. These students have been highly successful in graduating even if this programme has strict admission and graduation requirements comparable to engineering and science topics.

How the project was structured

The project was structured around several initiatives that were implemented and disseminated over several years. Figure 8.1 presents the broad vision of this project and its initiatives. This section will describe a specific initiative: ECL at UW – Design a Lab Facility for Hafumbure High School (Kenya). This initiative relied on another initiative that was ongoing in Kenya: ECL at MMUST – Build a Lab Facility for Hafumbure High School (Kenya). The interaction between these two initiatives exposed students in the two universities to challenges associated with geographically dispersed teamwork because Kenyan students had to collect field and material information and provide it to UW students, who used it to work collaboratively to design a lab for a rural high school that was going to be built by the local community in Kenya under the assistance of MMUST civil engineering students. This interaction was described in detail in a short YouTube documentary.[1] UW-based activities resulted in the development of a pilot course aimed at achieving an enhancement of interdisciplinary thinking, teamwork, and collaboration across CM, Arch, and CE students while delivering a set of design alternatives for our partner in Kenya. The next section describes how the instructors designed this pilot offering. Then information on student learning outcomes as well as a set of instructors' lessons learned is provided.

Issues encountered with the project

To enhance interdisciplinary thinking and learn how to collaborate with other disciplines, UW faculty decided to rely on a design-build course. At UW, design-build studio classes are often offered to a student audience that includes construction management students as well as students from design disciplines. This approach is easier to apply when students in the various programs follow a similar schedule. However, this is often not the case, so instructors' desire for pedagogical innovation often crashes against the complexity of fitting the new class into a fairly rigid programme schedule.

The authors encountered these scheduling issues when they initiated work to deliver a joint design-build course for CM, Arch, and CE students. The nature of the project called for a design-build studio, but the instructors found serious challenges in fitting a studio course into the busy schedule of construction management and civil engineering students on such short notice. At UW, students in the construction management and civil engineering programmes follow two separate curricula that are both built around a lecture-type schedule relying on the usual 3 to 4 contact hours per week. On the other hand, architecture students follow a schedule that is heavy on studio courses that rely on 6 to 12 contact hours per week. The main question for the instructors was "How to allow students in lecture-based curriculum programmes to participate in studio-like courses where new ways of experiencing teamwork and interdisciplinary thinking may be investigated?" To respond to this challenge, the instructors piloted a new course model that relied on team-level independent studies to be concluded with a design-build competition judged by a group of practitioners, which included a contractor, an architect, and a project manager experienced with Kenyan construction. This pilot offering was organised so that students worked in teams under independent study credit, and faculty coached the students as they worked through the problem. There were two faculty members who committed 3 to 5 hours per week on the project. There was also a research associate (RA) hired to organise and facilitate the class, and this RA spent between 5 and 10 hours per week. The faculty and RA time included planning presentations and invited guests, meeting and coaching the students, evaluating interim deliverables, and coordinating the communication with Kenya partners and industry guests.

Working in teams, students enrolled in this independent-study course were expected to integrate their architectural, engineering, and construction management knowledge to design and plan construction of an educational facility. As part of the assignment, teams had to produce an electronic building information model that would result in a set of detailed documents, including building plans, sections, exterior wall details, and material selection, electronic three-dimensional model of the facility, quantity take-off, and a project schedule.

Each team was comprised of four to five undergraduate and graduate students in architecture, construction management, and civil engineering. On each team, we had either a CM master student or an Arch/CM dual-degree student. Each team also had a civil engineering student who focused on structural design. The other team members were juniors and seniors in the architecture and construction management programs.

Throughout the quarter, we met as a "class" once per week, and CBE faculty and industry representatives provided short presentations on various topics. On weeks without presentations, instructors and guest industry coaches met with each team individually and provided coaching and feedback on their projects. Presentation topics included:

- How to successfully participate in a design-build competition;
- Challenges and lessons learned with international-cooperation projects;
- Rural education in Kenya: challenges and needs; and
- Availability of construction materials and methods in Kenya.

Since we were not meeting regularly throughout the week, to keep the teams on track, we gave them interim deliverables as listed in Table 8.1.

Table 8.1 Design definition submittal schedule

Due Date	Deliverable	Scale
April 12	Floor plans and site layout	1:100
April 19	Floor plans, 2 building section, and elevations illustrating all sides of your building	1:50
April 26	Structural plans and sections	1:10

The final deliverables included a final notebook (i.e., digital pdf and two bound 11 × 17 colour copies), a final presentation (i.e., digital pdf/PowerPoint and eight summary sheets for reviewers, as well as any other supporting documentation teams wished to provide, such as calculations and estimates). The final notebook was organised into sections and included at minimum the following information: Design, Conceptual Plan Sequence, Quantity Estimate, and Sustainability. We asked the students to emphasise the integration of the project across all of the dimensions and disciplines. See Figures 8.2, 8.3, and 8.4 for examples from the final notebooks.

a

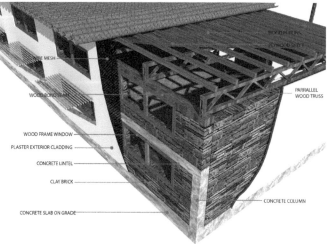

b

Figure 8.2a and 8.2b Overview of design concept and detail of wall construction by Team 1

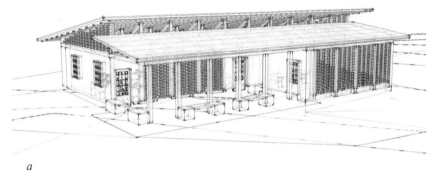

a

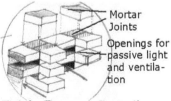

Mortar Joints

Openings for passive light and ventilation

Brick Screen Detail

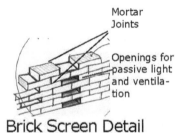

Mortar Joints

Openings for passive light and ventilation

Brick Screen Detail

Plan Diagram

Section Diagram

b

Figure 8.3a and 8.3b Schematic overview of design concept by Team 2

a

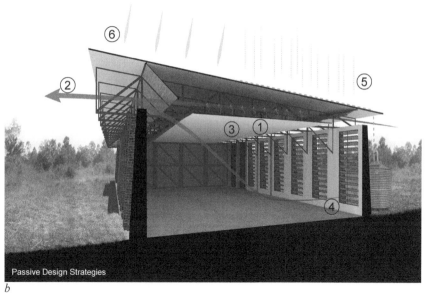

b

Figure 8.4a and 8.4b Overview of design concept and passive cooling strategy by Team 3. In Figure 8.4b, note the following: 1) 100% all natural daylighting, 2) Space left for maximum ventilation when desired, 3) Suspended radiant barrier and cloth protects from radiant heat exposure, 4) Louvered reed screens allow for airflow, shading, and security, 5) Rain collection for community and students, 6) Possibility for solar array to power computers

Findings of the project

In general, the pilot offering was successful in many ways. First, the student engagement with this course was outstanding. Once each team completed its presentation, the judges commented that they were "blown away" by the amount of work the students put into the projects. Judges found all three projects to be very strong and thought they could actually be constructed with only minor modifications to the design. The students themselves reported that they enjoyed the class and relished the opportunity to work on such a unique design problem, as well as work with students from different departments. The civil engineering students, in particular, said that it was one of the best experiences they had had in their program because they applied their new-found engineering skills to a real world project and learned a great deal about the other disciplines on the team. Based on the final design and construction concepts, the course also enhanced interdisciplinary thinking. Students from the three disciplines had to negotiate some design and construction aspects to generate the most innovative and constructible concept. Figure 8.5 includes excerpts of student final deliverables that exemplify outcomes of interdisciplinary work.

For instance, students were informed that Kenyan workforces mostly rely on informal labourers. Team 3 decided to incorporate this construction-specific issue into their work.

Their proposed solution was a notebook that included an Ikea-type manual with a list of needed tools linked to a detailed sequence of tasks extracted from the construction schedule (Figure 8.4a). Team 3 also strongly relied on the CE student to verify the structural feasibility of the team's truss design while attempting to identify the optimum design between structural, constructability, and passive ventilation requirements (Figures 8.4b and 8.3). Similarly, Team 2 explored different design concepts, trying to optimize natural lighting while pursuing a high level of integration of their design concept with existing buildings in the Kenyan campus and with local construction techniques (Figure 8.4c). Last, Team 1 strongly relied

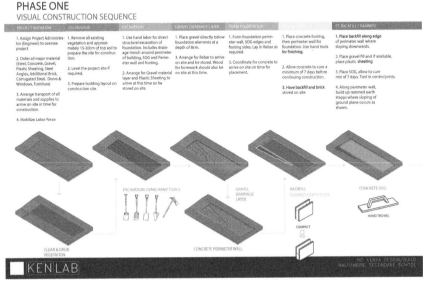

a

Figure 8.5a, 8.5b, 8.5c and 8.5d Examples of interdisciplinary work by teams: (a) sequencing and tool needs – Team 3; (b) structural analysis of trusses – Team 3; (c) daylighting – Team 2; (d) sequencing and constructability – Team 1

STRUCTURAL ANALYSIS

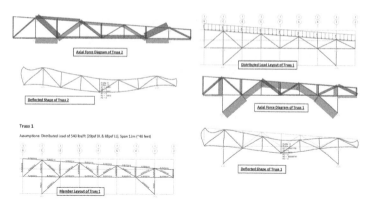

b

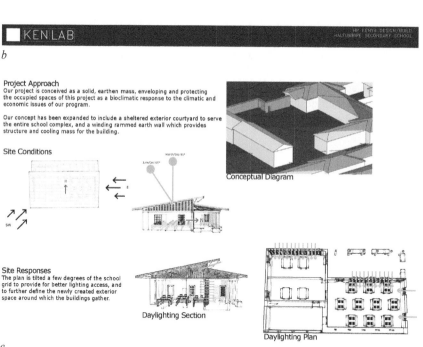

c

Figure 8.5a, 8.5b, 8.5c and 8.5d (Continued)

on BIM models to facilitate the interdisciplinary thinking as exemplified in their 3D schedule (Figure 8.4d).

The instructors also learned from this pilot offering. The main learning was a framework to develop an elective course for the CM and architecture curricula with the option for civil engineering students to seek admission in this course and

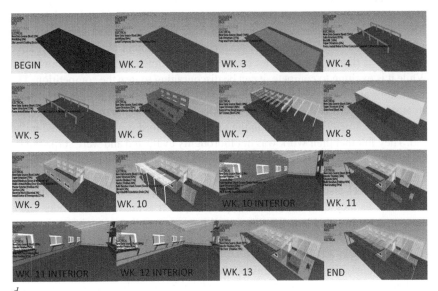

d

Figure 8.5a, 8.5b, 8.5c and 8.5d (Continued)

obtain elective credits. Our lessons learned relate to course scheduling, operational aspects, drawbacks from competitive environment, and handling teamwork as the prevalent part of a course.

In terms of course scheduling, the pilot offering worked well. Even if only 3 hours per week were allocated for in-class work, each team was well staffed, with a very diverse and comprehensive knowledge pool, so they were able to deliver high-quality work while needing a level of interaction that was adequate to a 3-credit teaching load. A contributing factor to this success was that each team included some students experienced in team-based studios who were able to provide internal guidance. However, if this offering would develop into an undergraduate elective course for design and construction students, we may expect that a lesser amount of experience may be available to student teams due to a less homogeneous cohort. As a result, teams may need more in-class guidance, especially for the design phase. This would mandate some changes in order to handle the course scheduling issues.

The pilot offering was somewhat plagued by communication issues with our partner in Kenya. While we were able to receive enough information to initiate the design process, this information flow was not continuous throughout the process, resulting in some unneeded stress for the instructor team and the students. Especially during the generation of design concepts, our partners were not responsive in terms of feedback and responses to the design teams' questions. Similarly, feedback was not provided on the draft concepts that were submitted for the client's feedback well in advance to the final competition.

Another lesson concerns the decision of running this course as a competition. In construction management education, researchers (Bigelow, Gilck and Aragon, 2013) found competitions to be very effective tools for student engagement in which the positive effects such as leadership development outweighed negative ones such as time. Still, this pilot course has also shown some of the drawbacks. While it is exciting to compete, it can be discouraging to lose the competition. The instructors and jurors jointly concluded that the competition seemed to inspire great student work and enthusiastic participation in the assignment; however, the instructors learned that students in the team placed third were disheartened in the end. They felt that their work was not thoroughly evaluated and were surprised by the results. While this is representative of the competitive nature of our industry and may constitute a learning opportunity in itself, it seems that the students who do not win feel discouraged and often do not have as positive a reflection after the course as they did during the competition. So it is possible to wonder about the long-term learning benefits of competitions versus other types of motivations for working together – such as public service and community outreach (e.g. Clevenger and Ozbek, 2013).

As previously mentioned, teamwork was another challenging aspect of this curriculum. Throughout the quarter, we learned that because this was an independent study, we had less of a chance as faculty to intercede or coach the team through team tensions. However, the students were able to work out tensions on their own. While conflict in group assignments can be seen as a challenge and distraction, it can also be seen as a learning opportunity in which each student experientially learns to deal with others whom they find difficult to work with.

Overall, the pilot offering worked fine, but as instructors, we need to create space where students can feel successful in their project while at the same time reflecting on what they learned from challenges along the way.

Conclusions

The pilot offering suggested that this course model could be replicated and improved once some of the lessons learned are taken into account. We envision that at least three replication paths are feasible depending on the program. First, utilizing this approach to offer a joint capstone course for students in the different design and construction programs sounds intriguing and challenging at the same time. Several design and construction schools already rely on team-based capstone courses. A team-based interdisciplinary and integrated design/construction capstone would prepare students for the industry environment and provide a bridge between academic and industry life. However, this approach adds significant complexity to the delivery of a capstone course and should not be considered if the student cohorts are not already familiar with teamwork and/or studios and have not been exposed to some level of interdisciplinary thinking throughout their previous coursework. A second path would be to include this course as an elective in the senior year before their capstone experience. At that time, students would build on their teamwork experience through previous studios or class projects to become familiar with interdisciplinary thinking before undertaking their capstone. This "softer" approach would reduce the stress of having a large interdisciplinary cohort, as it would be for a capstone project. However, it would still rely on students to be somehow familiar with teamwork. Last, this course could also be offered earlier in the curriculum, maybe as a conclusion of their junior year, and be used to prepare students for the same level of teamwork and interdisciplinary thinking. At that time, students may have little to none familiarity with teamwork, so we envision that a hybrid studio/independent study could be the solution. Under this approach, a traditional studio of 6 hours per week could be scheduled for architecture students, while the independent study course for CM and CE students could overlap for 50% of the time to allow in-class coordination and collaboration among disciplines. The extra studio time would allow students in architecture to refine their design and interact with the instructors. While involvement from CM and CE faculty and students would bring interdisciplinary experiences, balancing course loads and expectations from the various programs may be challenging and can be achieved by seeking support by leadership at the participating units. Last, if the primary goal of this course is to create a positive environment for cross-disciplinary collaboration, we have found that the public service aspect of this project provides a clear focus for the students – to help people by improving their built environment. The reality of helping people is a huge motivator for Millennial students and is an incentive for them to stay focused on the higher goals of the project, giving more purpose to their work than a synthetic project could. However, this objective can be achieved only if the partner/client organisation is responsive and is willing to give back to the students in the form of feedback and comments as much as it receives from them. While being challenging, this issue can be resolved through a strategic partnership with a nonprofit organisation whose mission is aligned to the public-service component of the project to be designed and built. While not totally successful in the pilot, the authors were able to create a cushion between the students and the clients so that lacking feedback from the client could be compensated by more feedback by the instructors and industry coaches.

Does access to technology facilitate experiential and contextual learning?

The Safety Inspector *video game*

A parallel project to the ECL+VI programme, led by Dr. Ken-Yu Lin at the University of Washington, is aiming to prove that technology and virtual reality support experiential and contextual learning in a way that is superior to traditional learning and testing methods. This involves collaboration with several researchers from different U.S. academic institutions to create and validate a video game called *Safety Inspector*, which simulates different situations in which a student must be able to recall the Occupational Safety and Health Administration (OSHA)'s best practices and regulations for job site safety (Liaw *et al.*, 2012).

The principal investigators believe this type of assessment is preferable for a student in the built environment discipline, because a written test cannot replicate a construction site environment, while the game environment simulates job site conditions.

Safety Inspector is based on a commercial game engine called Torque 3D (InstantAction, Las Vegas, NE, USA). The game logic resembles a typical first-person action game. The players, impersonating a safety inspector, can freely explore the game environment (i.e. a construction site) and encounter situations that may or may not be job hazards. Protruding nails on plywood, using ladders on scaffolds, ladder not being set up correctly, materials being stacked to high and too close to unprotected roof edges, and workers not wearing proper personal protective equipment are some of the example hazards being modelled. The game environment consists of three separated areas. Each area corresponds to a specific construction phase such as earthwork and foundation, superstructure, and exterior enclosure. Thus, the players can experience a great diversity of hazards and violations that may happen during a typical construction process.

Worker and equipment appearance and movements are modelled to represent in detail both static (e.g., nails protruding on plywood) and dynamic (e.g. two workers climbing the same ladder at the same time) types of hazard and violations. To increase the game's realism, objects generally present in a construction site (e.g., temporary offices, cars, power lines, and trees) and background buildings and roads are also reproduced in the game.

The players can interact with the game using both the touchpad and the keyboard. In particular, the pointer is used to aim and select the possible hazards. When an item is selected, the game provides a confirmation message telling whether the worker, tool, piece of equipment, or condition is a violation. Further, the players are awarded with game points when they identify a violation. The amount of points awarded for a violation is based on the level of knowledge necessary for identifying the violation.

The implementation of Safety Inspector *at the University of Washington*

Construction safety is a required course for most undergraduate and/or graduate programs in construction engineering and management. Examples of the topics are construction safety rules and regulations, hazard recognition techniques, how to control and minimize hazards, and how to effectively implement personal protective equipment. The UW offers a construction safety class for undergraduate CM students in their senior year. Usually, the students perform a series of activities designed to learn and practice hazard-recognition skills during the quarter. Then, at the end of the quarter, they take a paper-based exam to prove their proficiency in recognizing hazards and violations. By providing a comprehensive hazard-recognition challenge, *Safety Inspector* is designed to capture learning performances that can be hardly measured by traditional assessment techniques, such as a paper-based exam. Thus, the main purpose of the research project is to assess if the video game could provide a better learning assessment than a traditional paper-based exam.

Students were engaged to experience and help evaluate the game by Dr. Lin, the instructor of the class; 39 of the 45 enrolled students during the autumn quarter of 2011 consented to participate in the study. *Safety Inspector* was presented to the safety class students at the end of the quarter along with the paper-based exam. The students were introduced to the game mission and learned how to play the game. Then they took the game-based test by playing the game for 10 minutes. In particular, 22 HP EliteBook Tablet PCs were provided by the UW HP Catalyst program to allow the students to learn the video game and take the game-based test. After completing the game-based test, the students took the paper-based exam. The scores on both test modes were used to validate the game-based test. To provide a comprehensive assessment of the game, students also filled out a questionnaire designed to capture test mode preferences, video game experience, and opinions about the game-based test (Liaw *et al.*, 2012).

How the project was structured

Two main sponsors support this research project: the National Science Foundation (NSF) by financing the development of the video game and the UW HP Catalyst Program by providing the laptop computers necessary to validate the video game. Further, several academic institutions other than the UW CM Department were involved in the development of the video game, including the Department of Education at the University of Washington; the Civil, Architectural and Environmental Engineering Department at the University of Texas at Austin; the Rinker School of Building Construction at the University of Florida; the Charles W. Durham School of Architectural Engineering and Construction at the University of Nebraska–Lincoln; and the Department of Architecture Engineering at Ewha Womans University in South Korea.

The validation of the game-based test was implemented in a construction safety course that is a required course in the UW Construction Management bachelor program. This program is accredited by the American Council for Construction Education (ACCE), which requires the teaching of construction safety among the Core Subject Matters. Topical content to be covered in construction safety classes includes (a) safe practices, (b) mandatory procedures, training, records, and maintenance, and (c) compliance, inspection, and penalties (ACCE, 2006). While experimenting with the use of *Safety Inspector*, Dr. Lin had to fulfil these requirements. As a result, the amount of in-class time dedicated to the experimentation was constrained by other competing objectives. In addition, the proposed activities involved human subjects and, therefore, were required to undergo review by the UW Institutional Review Board.

Conclusions

The test results indicated that the game-based and paper-based tests can be comparable under certain conditions to yield statistically correlated test scores. For the video game alone, students responded positively regarding the game's features and values in support of their learning. Certain type of hazard-recognition challenges might need more innovative ways of task construction and presentation, as they not work as well in the current version of the game-based test which require a game player to look up (e.g. overhead power line) or down (e.g. uncapped rebar) in the virtual game environment. When the hazard-recognition challenge demands the representation of a material's actual texture in colour, neither of the two test modes is satisfactory. Findings will be further studied in the next research stage, and the authors hope that their experiment strategies, planning, and outcomes could benefit the AEC community who are interested in discipline-specific education innovations.

Note

1   www.youtube.com/watch?v=tsPhHchhOog

References

ACCE. (2006). *Document 102: Manual for the Preparation of the Self-Evaluation Study.* US: San Antonio, TX: ACCE (American Council for Construction Engineering), http://acce-hq.org/documents/document102120111.pdf.

Bigelow, B.F., Glick, S., and Aragon, A. (2013). Participation in Construction Management Student Competitions: Perceived Positive and Negative Effects. *International Journal of Construction Education and Research*, 9(4), 272–287.

Clevenger, C.M., and Ozbek, M.E. (2013). Teaching Sustainability through Service-Learning in Construction Education. *International Journal of Construction Education and Research*, 9(1), 3–18.

Liaw, Y.L., Lin, K.Y., Li, M., and Chi, N.W. (2012). Learning assessment strategies for an educational construction safety game (pp. 2091–2100). *Proceedings of Construction Research Congress 2012*, May 21–24, West Lafayette, IN.

9 Successful Education of professionals for supporting future BIM implementation within the Architecture Engineering Construction context

Farzad Pour Rahimian, Jack Steven Goulding and Tomasz Arciszewski

Introduction

Scientific advances and innovative technologies in Architecture, Engineering and Construction (AEC) projects are shaping the way in which decisions are made. These decisions are forming the governing foundations for determining and delivering progressive changes in order to address issues pertaining to society, knowledge, economy and people. These central tenets help support, underpin and drive societal drivers. However, whilst the AEC sector as a whole (in the UK) has been fragmented (Latham, 1994; Egan, 1998; Wolstenholme, 2009), the consequence of this has hindered progress in the use of such innovative technologies (Pour Rahimian *et al.*, 2011). That being said, from a United Kingdom (UK) perspective, there has been a resurgence to address these challenges in order to exploit the full potential of such technologies. In doing so, the UK government mandated the implementation of building information management (BIM) Level 2 in all public projects from 2016. This required AEC firms to use exchangeable digital building models amongst project parties to enable 2D/3D spatial compliance with British Standard BS1192:2007.

This approach has been mandated to optimise project costs and labour by eliminating redundant efforts and parallel production of multiple design solutions. Such an integrated design and implementation platform entails a high degree of automation throughout the whole AEC practice, which is currently hindered at very early conceptual design and planning phases, since advanced visualisation and modelling technologies are employed only at the detailed design stages. As such, many pioneering research institutes are now concerned with leveraging ultimate adoption of BIM through bridging this gap by promoting the subject of IT integrated design and construction, or in other words Integrated AEC, which originated at Stanford University.

With this innovative approach to design, the new generation of AEC professionals needs to be educated how to develop not only traditional or routine projects but also projects incorporating novel designs and construction processes. These professionals need to be creative, and be able to develop unknown (or unproven) solutions which are feasible, surprising, and potentially patentable. Currently, AEC professionals are no longer being seen as leaders or innovators but more as followers – using deductive problem solving rather than seeking opportunities, using their creativity and developing inventions. This resonates with thinking derived from innovation literature (Akintoye, Goulding and Zawdie, 2012; Elmualim and Gilder, 2014). As a result, designers and engineers in particular have seemingly lost their ability to innovate. This is partly attributable to 'inappropriate' education that has historically focussed on production rather than creativity. This is just the opposite of what happened in the 19th and early 20th centuries, when designers and engineers were seen as the 'true drivers' of change. During this time, high-level education was aligned to incentives (e.g. the highest salary rates), which helped design and engineering schools attract the most talented students, and these graduates were capable of meeting all technological and socio-cultural challenges of the quickly expanding societies (Arciszewski, 2006; Arciszewski and Rebolj, 2008; Arciszewski and Harrison, 2010a, 2010b). For instance, the construction of some monumental buildings during this period in history created not only technological solutions, but also cultural revolutions, leading to a fundamental change in the way design and engineering were perceived.

This research posits that creativity has increasingly been underrepresented, and as such, needs to be revisited, especially in a rapidly evolving technologically driven world. For example, such challenges now include environmental and sustainability demands, increased levels of safety compliance, enhanced security issues and whole-life demands (energy, maintenance etc.). Whilst it could be argued that some of these challenges extend beyond the AEC domain *per se*, it is important to identify the key promoters and inhibitors of engineering creativity. In doing so, the profession as a whole will benefit from a new cogent way of embedding creativity into solutions, the result of which will not only benefit society but also help inspire future AEC successors to follow this approach. Any changes, particularly those related to the ways that AEC students are educated, are extremely difficult, mostly because of the vector of psychological inertia (Altshuller, 1984a) in action. This phenomenon refers to a natural tendency of individuals and communities to resist any changes, thereby delaying progress as much as possible. This is also influenced by the way in which the instructors were originally educated (mostly as highly sophisticated analysts), as this has a significant impact on the way they want to teach students. Cognisant of this, it is important to recognise the need to apply a complex systems approach to analyse the impact of this in order to address the current situation.

This chapter presents design and engineering leadership as three interrelated abilities: (1) to develop a vision, (2) to transform it into a strategy and (3) to implement it. The key to leadership is the ability to develop feasible ideas or concepts (e.g. a new type of engineering system or construction process) using a set of abilities (traits) required to implement them (as opposed to using existing concepts to perform typical/routine work), in particular, the development of a vision similar to conceptual design, especially to inventive design. In both cases a new idea, or a concept of an engineering system, needs to be developed. This is the area of activities in which creativity or abductive generation of new ideas takes place. This position is proffered, as historically, 'followers' have been seen to create stagnation, producing what has been called "Vector of Psychological Inertia" (Altshuller, 1984b), or fixation (Youmans and Arciszewski, 2014). This psychological phenomenon therefore tends to make change and progress more difficult, and in some cases often even prevents it. The emphasis therefore is to consider the development of leaders (not followers) in order to minimise the negative impact of the vector of psychological inertia.

Building upon the principles of the *Theory of Successful Intelligence* (Sternberg, 1985, 1996, 1997), this chapter describes 'success' as a relative concept, which is defined by a given person in relation to the socio-cultural context and personal desires. This study therefore posits that there is a need to develop a new paradigm that recognises the importance of both analytical and creative works. Given this, this research defines analyst learners as the people who use rote learning and deduction, eventually induction, as opposed to creative people who use also abduction for reasoning. This approach extends learning capability beyond the learners' cognitive capability. Relying on the principles of *Theory of Successful Intelligence* (Sternberg, 1985, 1996, 1997), positive psychology (Schueller, 2012) and appreciative intelligence (Barrett and Fry, 2008), this chapter asserts that by using the 'right' methodologies and media, general principles of creative work could be translated into an explicit knowledge form and become part of a body of knowledge, hence enabling the "Successful Departments" (Arciszewski, 2009) to teach learners the "Creative Intelligence" and "Appreciative Intelligence". In this context, the potential of utilising advanced visualisation tools such as immersive game-like virtual reality interfaces is deemed vital – especially for augmenting analytical and parametric thinking capacity to intuitive idea generation (which could both be supported by these interfaces).

Learning and pedagogy

The AEC sector engages a wide range of diverse stakeholders, each with specific skill sets, learning requirements and contextual boundaries. Moreover, it also needs to be recognised that individuals (even within a context silo) have specific learner needs and disposition to learn in a certain way. Thus, it is important to understand how learners learn, as this ultimately impinges on the overall success of the learning process. Acknowledging this, a variety of models have attempted to define and characterise learning styles (Coffield *et al.*, 2004; Karagiannidis and Sampson, 2004; Goulding and Khuzzan, 2014) in which learning has proven to be more effective where the instructional process supports the various learning styles of learners (Kolb, 1984; Kim and Chris, 2001). Given the cornucopia of stakeholders within AEC, it is particularly important to appreciate that a learner's learning experience should be as personalised as possible (Vincent and Ross, 2001), as a 'one-size-fits-all' approach is generally ineffective (Watson and Hardaker, 2005). Thus, there is a need for learning instructors/trainers to take learning styles into account, especially where new technology developments (e.g.

BIM, game-like technologies, immersive technology etc.) offer new opportunities. This thinking, particularly using adaptive technologies (which incorporate behavioural and attitude measures), is increasingly gaining momentum.

Education, training and pedagogical development embrace theoretical and applied research which draws upon the theory of social sciences (Watkins and Marsick, 1992; Klimecki and Lassleben, 1998; McAdam, Leitch and Harrison, 1998; Pemberton and Stonehouse, 2000). Given that this chapter acknowledges the need to apply successful education to AEC professionals wishing to engage with technology-driven solutions (such as BIM), it uniquely aligns itself to both social sciences theory and behavioural science theory; as it fervently proffers that there is a real need to move away from traditional education and training delivery approaches. Research has shown that a match between learning environments and learners' learning styles can enhance learners' performance, motivation and efficiency (O'Brien, 1989; Oxford and Ehrman, 1993; Brown, 1994; Chang and Cox, 1995, Naoum and Hackman, 1996; Kumaraswamy, 1997; Buch and Bartley, 2002; Karagiannidis and Sampson, 2004). This resonates with following section, which identifies how the *Theory of Successful Intelligence* relates to context, place and instructional goals.

Theory of Successful Intelligence

The *Theory of Successful Intelligence* (Sternberg, 1985, 1996, 1997) is a major step toward understanding how individuals' abilities are interrelated with their life success. In the context of design and engineering education, this theory presents a new understanding of how education can be conceptualised, designed and delivered. Through this theory, successfully intelligent people are defined as those able to achieve their goals by leveraging their strengths, compensating for their weaknesses and adapting to and shaping and selecting environments that will facilitate their success. This theory is underpinned by three fundamental pillars:

1   Successful intelligence can be learned;
2   Successful intelligence is a combination of three independently acquirable abilities, namely practical intelligence, analytical intelligence, creative intelligence;
3   Successful intelligence is dynamic; both the criteria of success and the abilities the individual employs (i.e. the relative combination of the three intelligences) to achieve success may change during one's lifetime.

In accordance with this theory, practical intelligence is an ability to solve simple, everyday problems, and this is done using easily available knowledge and heuristics. Abilities to open a door or to ride a bus are good examples of practical intelligence. Analytical intelligence is an ability to solve analytical problems, and that requires using deductive skills and utilising existing knowledge (for example, analysis of traffic flow, numerical optimisation or planning a typical construction process, etc.). Analytical intelligence is acquired through the combination of rote learning and learning deductive skills. Analytical intelligence alone is what traditional IQ tests measure. In addition, traditional engineering education emphasises analytical intelligence almost entirely. However, the *Theory of Successful Intelligence* stipulates that a balance of the three pillars of intelligence is absolutely necessary for life success, including professional success.

In the AEC context, creative intelligence is the ability to solve inventive problems, which require abductive skills and obviously the use of existing knowledge. Solving such problems requires development of unknown solutions or ideas, for example development of a new type of a wind bracing system in a tall building or a new type of a tunnel. Creative intelligence is acquired through the combination of rote learning with learning of both deductive and abductive skills.

Successful education

Successful Education (Arciszewski, 2009) is a new paradigm in design and engineering education. This paradigm was inspired by the latest developments in modern cognitive psychology, especially by the *Theory of Successful Intelligence* (Sternberg, 1985, 1996, 1997). This paradigm has also been strongly influenced by a new understanding of historical and social mechanisms behind the emergence of the Renaissance, including the Medici Effect (Johansson, 2004) and the Da Vinci Principles (Gelb, 1998, 1999, 2004). Arciszewski (2009) argued that principles were particularly important because they provided a synthesis of all attitudes practiced by Da Vinci and by the other great Renaissance engineers.

In this paradigm, the key concept is "Successful Designers and Engineers", and it describes the designers and engineers who have not only acquired as students the necessary and sufficient body of knowledge to practice engineering but also learned Successful Intelligence, including its three components, practical, analytical and creative intelligence. Such graduates are prepared to not only undertake any kind of routine work but, if necessary, to become inventors and leaders, since in both cases the key to success is an ability to develop new ideas.

In Table 9.1, Successful Education is compared with a past design and engineering education paradigm, called the 'master–apprentice paradigm', and the present one, named by the authors is the 'scientific paradigm'. The comparison is done from the perspective of the *Theory of Successful Intelligence* and of its three main components. In this context, only Successful Education is complete since only it addresses all three components of Successful Intelligence and consequently creates an opportunity to educate successful engineers.

Table 9.1 Comparison of teaching paradigms

Teaching Paradigm	Practical Intelligence	Analytical Intelligence	Creative Intelligence
Master–Apprentice	Yes		Yes
Scientific	Yes	Yes	
Successful Education	Yes	Yes	Yes

Successful Education requires not only a new understanding of design and engineering education priorities and several new or modified courses, it also requires a complex environment, called 'Successful Department', which will enable and stimulate the creation of successful engineers. A modern Medici Effect and the resulting intersection of ideas are crucial for the learning process. Therefore, they require a revolutionised environment (in terms of intellectual and technological structures) which is completely different from the current 'look' of so many design and engineering departments. In essence, there are four major components of a Successful Department, namely courses, instructors, physical environment and ambience (Arciszewski, 2009). This is aligned with Salama's (2008) *'Integrating Knowledge in Design Education'* theory, which argues that a responsive architectural design pedagogy giving credit to socio-cultural and environmental needs can enable future architects to create liveable environments.

Traditional, analytical courses are absolutely necessary for the future of successful engineers, although they are grossly insufficient for them. They require additional courses on inventive design and engineering, focused on the emerging science of inventive problem solving. For the best results, such courses could/ should be offered to students through their entire period of studies. A single course for seniors (the present practice at George Mason University) is a step in the right direction, but it comes too late to impact learning in other courses and to transform students into successful engineers. A much better solution is a sequence of several courses, even if the total number of credit hours is the same.

Instructors are the key component of a Successful Department. Usually, faculty in academic units are surprisingly similar in many aspects (birds of a feather flock together) despite all efforts to create diversity, which is often imposed only for political reasons. A Successful Department requires, however, true diversity, which may be described as 'balanced intersection'. This term is understood as a selection of instructors resulting in a department in which cultural backgrounds of instructors are strongly differentiated; they represent both applied and fundamental research, have experience in analytical and exploratory research and represent various thinking styles.

A physical environment creates a framework for learning and also sends a message about the nature of a given academic unit (Hou and Ji, 2010). An ideal urban design for a Successful Department should be based on the concept of the agora as an ideal form stimulating human interactions through complex socio-psychological mechanisms. Such an urban complex should have several buildings arranged around the central square/agora. A building should be dedicated to teaching practical intelligence and designed with all kinds of testing laboratories and workshops. Another building should be dedicated to teaching analytical intelligence, and it should have various computer laboratories. A third building, 'Inventors' Heaven', a must, should be dedicated to teaching creative intelligence with appropriately selected laboratories and workshops specifically designed for teams working on their inventive challenges. Finally, there should be an administrative building for faculty and classrooms.

A Successful Department would never be fully effective without a 'proper ambience'. In this case, ambience is understood as a multisensory experience that positively affects students, faculty and staff, helping them to learn or teach in the best way to create successful engineers. Ambience obviously has an emotional dimension, which distinguishes it from a traditional department. Ambience is a reflection of people's perception of an environment surrounding them and can be carefully created in such a way as to contribute successful designers and engineers. Arciszewski (2009) discussed various components of ambience in a Successful Department, for example guiding principles and stories, colours, music, art, various activities and even the proper lighting in the Successful Department.

Building upon the theoretical bases discussed in the theory of Successful Education (Arciszewski, 2009), this chapter highlights the potentials of the advanced IT interfaces for leveraging all four components of such a Successful Department. The chapter particularly suggests use of advanced game-like virtual workspaces in order to leverage education of successful designers and engineers for the AEC professions.

Games and virtual reality in construction engineering education

The nature and complexity of communication mechanisms within AEC projects has changed significantly over the last 10 years, especially the modus operandi and integration with core business operations. This has been reflected through the increased prevalence, use and deployment of web-based project collaboration technologies and project extranets. Within the AEC sector, information and communications technology (ICT) has revolutionised production and design (Cera *et al.*, 2002), which has led to dramatic changes in terms of labour and skills (Fruchter, 1998). However, it is also important to acknowledge that the capabilities of such applications (and implementation thereof) in predicting the cost and performance of optimal design proposals (Petric *et al.*, 2002) should enable design engineers to compare the quality of any one tentative solution against the quality of previous solutions. This was further reinforced by Goulding, Pour Rahimian and Wang (2014) regarding the ability to experiment and experience decisions in a 'cyber-safe' environment in order to mitigate or reduce risks prior to construction. It is therefore crucial for the AEC industry to employ cutting-edge ICT technologies to issues related to organisational management and decision making (Friedman, 2005). Furthermore, whilst advocates note that these have helped to resolve some of the aforementioned challenges, Pour Rahimian *et al.* (2011) noted that project teams are still facing real and significant problems and challenges regarding heterogeneous systems faced by project teams using project extranets. In this essence, the problem here is that the industry is experiencing confusion as to how to manage project information in order to support decision-making processes. This is the point at which Fruchter (2004) suggested the digital integration of the whole data creation, retrieval and management system within the building industry in order to prevent tacit knowledge loss and miscommunication among various parties from different disciplines. In this respect, recent innovation in virtual reality (VR) technologies and AEC decision-support toolkits have now matured, enabling telepresence engagement to occur through integrated collaborative environments. Several opportunities are now available, including significantly improved immersive interactivity with haptic support that can enhance users' engagement and interaction.

Employing cutting-edge ICT tools is also expected to leverage training systems within the AEC sector (Fruchter, 1998), as the implementation of effective training could impact the whole industry by addressing and fulfilling the needs of the different stakeholders in the industry. In this respect, advanced ICT systems are expected to address the shortcomings of 'typical' learning models that often provide the trainees with only general instructions (Laird, 2003) and issues associated with unaffordable costs of the 'traditional' on-the-job trainings (Clarke and Wall, 1998). Therefore, new ICT advancements that incorporate innovative proactive experiential learning approaches which link theory with practical experience using virtual reality interactive learning environments can be especially effective (Alshawi, Goulding and Nadim, 2007).

As an underpinning technology, VR has been defined as a 3D computer-generated alternative environment to be immersed in, for navigating around and interaction (Briggs, 1996), or a component of communication taking place in a 'synthetic' space, which embeds a human as its integral part (Regenbrecht and Donath, 1997). The definitions of VR systems usually include a computer capable of real-time animation, controlled by a set of wired gloves and a position tracker and using a head-mounted stereoscopic display as visual output. For instance, Regenbrecht and Donath (1997) defined the tangible components of VR as a congruent set of hardware and software, with actors within a three-dimensional or multidimensional input/output space, where actors can interact with other autonomous objects in real time. VR has also been defined as a simulated world, which comprises some computer-generated images conceived via head mounted eye goggles and wired clothing, thereby enabling the end users to interact in a realistic three-dimensional situation (Yoh, 2001).

Over the last 30 years, ICT systems have matured and enabled construction organisations to fundamentally restructure and enhance their core business functions. Sampaio and Henriques (2008) asserted that the main objective of using ICT in the construction field is supporting management of digital data, namely to convert, store, protect, process, transmit and securely retrieve datasets. They acknowledge the commencement of VR techniques as an important stepping stone for data integration in construction design and management, as they are capable of holding and presenting complete information about buildings (e.g. size, material, spatial relationships, mechanical and electrical utilities etc.) through a single output. Similarly, Zheng, Sun and Wang (2006) proposed the use of VR to reduce time and costs in product development and to enhance quality and flexibility for providing continuous computer support during development life cycle.

Early studies that incorporated VR into the design profession used it as an advanced visualisation medium. Since as early as 1990, VR has been widely used in the AEC industry, as it forms a natural medium for building design by providing 3D models, which can be manipulated in real time and used collaboratively to explore different stages of the construction process (Whyte, Bouchlaghem and Thorpe, 1998). It has also been used as a design application to provide collaborative visualisation for improving construction processes (Bouchlaghem *et al.*, 2005). However, expectations of VR have changed during the current decade. According to Sampaio and Henriques (2008), it is increasingly important to incorporate VR 3D visualisation and decision-support systems with interactive interfaces in order to perform real-time interactive visual exploration tasks. This thinking supports the position that a collaborative virtual environment is a 3D immersive space in which 3D models are linked to databases, which carry characteristics. This premise has also been followed through other lines of thought, especially in construction planning and management, by relating 3D models to time parameters in order to design 4D models (Fischer and Kunz, 2004), which are controlled through an interactive and multi-access database. In similar studies, 4D VR models have been used to improve many aspects and phases of construction projects by (1) developing and implementing applications for providing better communication among partners (Leinonen *et al.*, 2003), (2) supporting design creativity (Rahimian and Ibrahim, 2011), (3) introducing the construction plan to stakeholders (Khanzade, Fisher and Reed, 2007) and (4) following construction progress (Fischer, 2000).

With regards to education, Wellings and Levine (2010) posited that there was a need to redesign the current text-based lessons into collaborative and multidisciplinary problem-based materials expressly to take on board real-world problems and solutions. They argued that this was not possible unless immersive and interactive games were employed for improving trainees' engagement. Similarly, Thai *et al.* (2009) asserted that pedagogical digital games offered an intact opportunity to enhance engagement of trainees and revolutionise teaching and learning. ACS (2009) summarised the benefits of the emerging educational interactive immersive game environments: (1) annotated objects could provide deeper levels of knowledge on demand, (2) incorporating additional dimensions of subjects (nD), (3) supporting distance team collaboration, (4) leveraging equal opportunities by providing distance-learning opportunities and (5) simulated learning by modelling a process or interaction that closely imitates the real world in terms of outcomes.

VR applications and game engines are now increasingly being used in the teaching and learning of AEC. According to Zudilova-Seinstra, Adriaansen and van Liere (2009), VR as a teaching tool can contribute to the trainees' professional future by developing some learning activities beyond what is available in the conventional training systems. With respect to educational issues in the AEC industry, Sampaio *et al.* (2010) argued that the interaction with 3D geometric models can lead to active learner thoughts which seldom appear in conventional pedagogical conditions. Moreover, Juárez-Ramírez *et al.* (2009) asserted that when augmented with 3D modelling, VR could lead to better communication in the process of AEC training. However, VR training environments have arguably not yet fully reached the potential of reducing training time, providing a greater

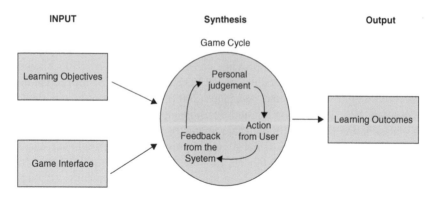

Figure 9.1 Educational game model input–synthesis–outcome (Garris, Ahlers and Driskell, 2002)

transfer of expert knowledge or supporting decision making. This was primarily down to the ways in which this technology was augmented. It is therefore argued that educational training tools need to 'engage' learners by putting them in the role of decision makers and 'pushing' them through challenges, enabling different ways of learning and thinking through frequent interaction and feedback and connections to the real world context (Goulding, Pour Rahimian and Wang, 2014). Furthermore, it is postulated that paring instructional content with game features could engage users more fully and help to achieve the desired instructional goals. In this respect, this study applied an input-process-output model (Garris, Ahlers and Driskell, 2002) of instructional games and learning to design an instructional programming, which incorporated certain features or characteristics from gaming technology, which trigger a cycle that includes user judgment or reactions, such as enjoyment or interest, user behaviour such as greater persistence or time on task and full learner feedback (Figure 9.1).

Cloud-based BIM in AEC areas

BIM is a model-based design process that adds value across the entire life cycle of the building project (Autodesk®, 2011). It is an intelligent integrating modelling tool for building design and construction which allows data sharing with all the stakeholders. It has been advocated that the key to implementation of BIM as the principal design delivery method is the ability of the various team members to easily share building information data during the design and construction processes. The information contained in a BIM model comes in various formats; thus it needs to be exchanged in an efficient way (Santos, 2009). Exchanging data can often be a challenge due to software incompatibility, different specifications, categorisation, format requirements and so on. However, addressing these issues can create interoperable systems that can help data modelling migrate between different teams with minimal data loss and with improved optimal accuracy (Fruchter, 1998).

Conventionally, prior to BIM adoption, architects created three-dimensional models merely for visualisation purposes and did not use them as data-rich intelligent models. In ideal practice, BIM models are even capable of supporting information exchange amongst various team members. In this form, BIM is not any more a production tool but a communication and social networking tool for designers. Succar (2011) explained various stages of BIM adaptation by introducing three major levels, namely modelling, collaboration and integration. The Australian Institute of Architects (2009) allocated the traditional production of two-dimensional documentation as stage zero in modelling implementation stages, thus rendering four capability stages in BIM altogether. Australian Institute of Architects (2009) proposed a model to divide these stages into two subdivisions, making each stage more specific in defining its capability. According to this model, BIM level one is defined as three-dimensional modelling (stage 1A) and intelligent modelling (stage 1B). Intelligent modelling also includes data attached to it, whilst three-dimensional modelling is merely for visualisation purposes only. BIM level two refers to the ability of two or more computer systems or software applications to exchange the format by following a standard and to make use of the information delivered. It is frequently defined as an interoperability system that allows the user to respond to the delivered model and customise it based on its requirements, specification and needs by utilising the nD's modelling concept – 4D = time, 5D = cost and 6D = facilities management.

BIM can be considered more than a representation tool or a means for developing a model or prototype to generate intelligent input. Additional benefits embrace several other issues, including facilitating the project teams to engage in innovative contractual relationships and new project delivery strategies. BIM level three offers an innovative way to excel in construction management. This new paradigm is known as integrated project delivery (IPD). Here, the goal was to create a team effort to increase good communication and team integration while working towards consensus. This is often called as the future of BIM. On this theme, Santos (2009) asserted that amongst all barriers for achieving this goal was the interoperability problems of BIM.

Interoperability refers to incompatibility between inter-products and software applications. Incompatibility means that vendors have created a solution to this by having a BIM model converted into a neutral object-based file format, a format that is not controlled by any particular vendor. Thus it can become a platform to exchange data. In essence, interoperability refers to the ability to exchange/share information between separate computer programs without any loss of content or meaning (Aranda-Mena and Wakefield, 2006). According to Succar (2009), interoperability is a linear workflow that allows the inability of simultaneous interdisciplinary changes to be shared in a single file-based sharing.

In the single operational file-based sharing model (Succar, 2009), once the building information model (1) is complete, it can be exported to the inter-operable model, BIModel (v1), to allow another process of modelling to be taken. This inter-operable model (v1) captures both geometry and properties of BIModel (1), thus facilitating the sharing of information. Then this inter-operable model (v1) will be imported to the BIModel (2) to allow a modelling process to take place;

this procedure will be repeated for another modelling process until the project is completed. The capability of this interoperability system allows BIM to take one further step to improve the interdisciplinary collaboration among the project team. This could be considered a stepping stone for web space–based platforms which are particularly beneficial for integrating visualisation components to give continuous related information sharing for the geographically dispersed end users.

One of the most referred-to industry (IEEE-1516) standards for large-scale modelling and simulation is the high-level architecture (HLA), which was originally introduced by the U.S. Department of Defense (Kuhl, Weatherly and Dahmann, 2000). Zhang *et al.* (2012) advocated for this system, as it could integrate various simulation applications, providing a standard architecture for interconnectivity, interoperability and reusability. Uygun, Öztemel and Kubat (2009) also posited that integrating various approaches and applications in computer simulations (using a unique framework, functional rules and common interfaces) could support flexible distributed simulations; moreover, doing so could contribute to the reduction of software costs by supporting the reuse of simulation models and providing an infrastructure for managing the runtime of the simulations.

From a similar point of view, Wang *et al.* (2014b) proposed a structured methodology for integration of augmented reality (AR) technology to BIM in order to overcome the issues related to limited sense of immersion and real-time communication of BIM within virtual environments. In a related attempt, Wang and Dunston (2013) developed a tangible mixed-reality interface to facilitate non co-located collaboration for problem solving and design error detection and Abrishami *et al.* (2013) proposed adopting generative algorithm into BIM to leverage its integration to conceptual design phases. Hou *et al.* (2013) developed another platform for controlling building components assembly procedures in order to improve accuracy and reduce errors. Wang *et al.* (2014a) adopted a more overarching approach and advocated the need for development of a computer-mediated remote collaborative design support system to leverage distributed cognition and help capture the non co-located team's knowledge, which is distributed in memories, facts, objects, individuals and tools.

This research extends the findings of previous studies in this area, with specific emphasis on supporting the decision-making process at the construction stages through the development of interactive and interoperable simulation platforms. The study provides a novel approach to support non co-located design teams using game-like VR environments blended to social sciences theory (social rules) and behavioural science theory (decision science/communication science). In essence, the aim of this study was to provide a flexible, interactive, safe learning environment for practising new working conditions with respect to offsite production (OSP) in general and open building manufacturing (OBM) in particular without the do-or-die consequences often faced on real construction projects. Hence, a VR interactive learning environment was sought which builds upon the multi-disciplinary practice-based training concept (Alshawi, Goulding and Nadim, 2007). In this context, the prototype aimed to enable disparate stakeholders with different professional specialisations to be exposed to the various aspects of OSP concepts.

This approach was adopted in order to help overcome the problem of 'compartmentation' of knowledge (Mole, 2003). Furthermore, the prototype had to be flexible enough to allow any-time-any-place learning so as not to be constrained to a particular place or time for learning to take place.

Case study

This section presents a developed web-based game-like VR construction site simulator (WBGVRSS). The primary aim of WBGVRSS was to embrace real-life issues facing off-site construction projects in order to appeal to professionals by engaging and challenging them to find real-life solutions to problems often encountered on site. Embryonic work by Goulding *et al.* (2012) established the philosophical underpinnings of this case study, and subsequent later developments are presented here for discussion. From the outset, a real construction project was used to map and govern construction processes – constructs, links and dependencies of which were embedded in a VR learning environment to replicate authenticity. In this context, the prototype learning simulator was designed specifically to allow 'things to go wrong' and hence to allow 'learning through experimentation' or 'learning by doing'. In this respect, although the 'scenes' within the simulator took place on a construction site, the target audience (learners) focussed primarily on construction professionals such as project managers, construction managers, architects, designers, commercial directors, suppliers, manufacturers and the like. Thus, the simulated construction site was used as the main domain through which all the unforeseen issues and problems (caused through upstream decisions, faulty work, weather, logistics etc.) could be enacted. The key learning impact areas acknowledged the importance and significance of these instances, placing emphasis on the decisions and subsequent implications on time, cost, resources and so on. Embedded learning scenarios were therefore user centric and progressive, especially where concatenated decisions had implications for succeeding events. Learning was therefore well planned and managed. All outcomes (of decisions) culminated in a formal report, which was used to reinforce learning through a debriefing session. This allowed learners to defend decisions and provide additional understanding, particularly with respect to mitigating circumstances. This reinforcement proved especially important for subsequent on-the-job learning. In this context, learning occurred through the following:

- Learner autonomy – to make all decisions;
- Interactivity – environment provides feedback on the decisions taken and their implications on the overall project (cost, time, resources, health and safety etc.);
- Reflection – users are able to defend decisions on the feedback provided and have the ability to identify means to avoid/mitigate potential problems in the future.

The main concept of the simulator was based on its ability to run scenarios through a VR environment to address predefined training objectives. In this respect, learning was designed to be driven by problems encountered in this environment, supported by a report critique on learners' choices, rationale and defence thereof. In accordance with these objectives, the WBGVRSS was designed and developed as an educational web-based simulation tool comprising both non-immersive and immersive pages for providing construction managers (and other disciplines) the opportunity to experience challenges of real-life AEC projects through simulated scenarios. In order to minimise interruption of the learners' reasoning process, the graphical user interface (GUI) was designed to be as simple and straightforward as possible with respect to data input. Thereby, the interface was designed to be accessible through any standard web browser to provide users with login account details and other criteria, for example selection of available construction sites, projects, contractors, equipment, scenarios and the like. All choices made by 'players' as well as their registration data were automatically recorded in a MySQL database, which was also accessible through the immersive application for project simulation. After completing the initial decision-making process through the interactive ASP.Net Web Forms, learners are able to commence the training session, starting with a 'walkthrough' to experience and appreciate the complexity of the project. At this stage, the application provides users with a summary of the project and contract and runs the simulation of the project within an immersive and interactive environment developed in Quest3D™ VR programming application programming interface (API).

Within the simulated Quest3D environment, the users are able to experience the outcomes of all decisions made. They are also challenged by unexpected events designed according to the selected scenario and are required to make decisions for dealing with these issues. The monitoring and communication tools are embedded in different parts of the main interface as well as the facilitated standard embedded virtual smartphone-type interface, which appears when required. The simulator ultimately records and tracks the users in the database and navigates to the conclusion page to reveal all scores of the user (together with the logic behind the marking procedure). Figure 9.2 illustrates a selection of the various functions available to the user of the simulator to fully interact with and retrieve information from the simulator during the VR simulation session. Further inclusion of the whole tree is considered for the exploitation phase.

Conclusion

BIM is now a prevalent factor in the procurement of AEC projects. However, these projects are increasingly becoming more complex, which not only requires new business processes and technological solutions to meet ever-increasing demands but also new skill sets. In particular, these business demands often require the conjoining of high-level skill sets to deliver the solutions needed. These skill sets are currently underrepresented and seldom engage the collective ethos needed to envelop creative thinking, through such approaches as pedagogical alignment

Figure 9.2 WBGVRSS simulation scenarios

and more specifically, Successful Intelligence in order to create new innovative solutions. It is therefore paramount that the industry as a whole engages the right type (and level) of skill sets and competence needed to meet these project requirements and business imperatives. Acknowledging this, it is also important that the causal drivers and influences associated with creativity and successful decision-making in global AEC teams are fully understood, engaged and supported. However, to do this requires a radical review of the way educational programmes and systems are designed and delivered. With respect to leveraging creativity and delivering innovation, this chapter reflected on the Renaissance period and the creativity-oriented learning/teaching paradigm called the master–apprentice paradigm, as opposed to the current analysis-focused science paradigm. From this, it introduced the *theory of Successful Intelligence* and its three pillars as an underpinning platform for educating the new generation of designers and engineers within AEC.

The "Successful Education" paradigm (Arciszewski, 2009) was presented as a new approach for educating AEC professionals. This included the concept of a new educational environment, the need for a new combination of courses that focus on teaching the three kinds of Successful Intelligence (in the context of the AEC sector) and specific guidelines of how to properly select instructors that are capable of implementing such approach. A proof-of-concept prototype using a web-based VR system and game-like cloud BIM platform for supporting integrated AEC projects was presented as an exemplar to demonstrate how the proposed approach could be implemented. This prototype simulator offers a risk-free environment in which learners can evaluate how decisions made affect business outputs. These decisions embraced (but are not confined to) design concerns, process challenges, communication, logistics and handling, supply chain management and so forth.

This chapter also proffered that enhanced engagement through an immersive project environment could lead to a better understanding (appreciation) of real-life AEC problems. This is particularly important when considering the need to procure 'value' through 'validated sustainable solutions'. Placing learners in a cyber-safe environment is one way of doing this, as it can leverage learners' cognitive processes to real-world issues without incurring the direct consequences of mistakes (which can be expensive – with far-reaching consequences). Finally, this premise and novel approach of applying game theory to non co-located design teams using game-like VR environments is an opportunity for industry reflection. This not only addresses the need to evaluate actor involvement in order to reveal new insight into AEC organisational behaviour but also the social constructs underpinning this (which often affect decision making). Advanced VR training and simulation tools are such exemplars on this metaphorical journey, which not only highlight the possibilities available but also emphasise the need to purposefully align pivotal drivers to specific learning outcomes. Future research in this area is likely to embrace the increased importance of pedagogy (learner styles/traits), as this is openly acknowledged as being particularly efficacious for delivering training material to specific learner types.

References

Abrishami, S., Goulding, J.S., Ganah, A., and Rahimian, F.P. (2013). Exploiting Modern Opportunities in AEC Industry: A Paradigm of Future Opportunities, in Anumba, C. (ed.), *AEI 2013: Building Solutions for Architectural Engineering – Proceedings of the 2013 Architectural Engineering National Conference* (pp. 320–332) State College, PA: United States.

ACS. (2009). 3D Learning and Virtual Worlds: An ACS: Expertise in ActionTM White Paper. www.trainingindustry.com/media/2043910/acs%203d%20worlds%20and%20virtual%20learning_whitepaper%20april%202009.pdf. [Accessed Date 22nd July 2011].

Akintoye, A., Goulding, J.S., and Zawdie, G. (eds.) (2012). *Construction Innovation and Process Improvement*. London: Wiley-Blackwell.

Alshawi, M., Goulding, J.S., and Nadim, W. (2007). Training and Education for Open Building Manufacturing: Closing the Skills Gap, in Kazi, A.S., Hannus, M. Boudjabeur, S. and Malon, A. (eds.), *Open Building Manufacturing: Core Concepts and Industrial Requirements* (pp. 153–172). Helsinki, Finland: ManuBuild in Collaboration with VTT – Technical Research Centre of Finland.

Altshuller, G. (1984a). *Creativity as an Exact Science*. New York: Gordon and Breach, Science Publishers, Inc.

Altshuller, H. (1984b). *Creativity as an Exact Science*. New York: Gordon and Breach, Science Publishers, Inc.

Aranda-Mena, G., and Wakefield, R. (2006). Interoperability of building information – myth or reality? *The 6th European Conference on Product and Process Modeling*, Valencia, Spain, 173–178.

Arciszewski, T. (2006). Civil Engineering Crisis. *ASCE Journal of Leadership and Management in Engineering*, 6(1), 26–30.

Arciszewski, T. (2009). *Successful Education: How to Educate Creative Engineers*. Fairfax, VA: Successful Education LLC.

Arciszewski, T., and Harrison, C. (2010a). Successful Civil Engineering Education. *ASCE Journal of Professional Issues in Engineering Education and Practice*, 136(1), 1–8.

Arciszewski, T., and Harrison, C. (2010b). Successful engineering education. *The International Conference on Computing in Civil and Building Engineering*, Nottingham, UK.

Arciszewski, T., and Rebolj, D. (2008). Civil Engineering Education: Coming Challenges. *International Journal of Design Science and Technology*, 14(1), 53–61.

Australian Institute of Architects. (2009). *National Building Information Modelling (BIM) Guidelines and Case Studies*. Brisbane, Australia: Construction Innovation.

Autodesk®. (2011). Realizing the Benefits of BIM, Autodesk® Building Information Modeling. http://images.autodesk.com/adsk/files/2011_realizing_bim_final.pdf. [Accessed 11/10/2016].

Barrett, F.J., and Fry, R.E. (2008). *Appreciative Inquiry: A Positive Approach to Building Cooperative Capacity*. Chagrin Falls, OH: Taos Institute.

Bouchlaghem, D., Shang, H., Whyte, J., and Ganah, A. (2005). Visualisation in Architecture, Engineering and Construction (AEC). *Automation in Construction*, 14(3), 287–295.

Buch, K., and Bartley, S. (2002). Learning Style and training Delivery Mode Preference. *Journal of Workplace Learning*, 14, 5–10.

Briggs, J.C. (1996, 09–01–1996). The Promise of Virtual Reality. *The Futurist*, 30–31.

Brown, A.L. (1994). The Advancement of Learning. *Educational Researcher*, 23, 4–12.

Cera, C.D., Reagali, W.C., Braude, I., Shapirstein, Y., and C. Foster. (2002). A Collaborative 3D Environment for Authoring Design Semantics. *Graphics in Advanced Computer-Aided Design*, 22(3), 43–55.

Chang, W.P., and Cox, R.P. (1995). A balance in construction education. *CIB W89 Conference on Construction/Building Education and Research Beyond 2000*, Orlando, 235–242.

Clarke, L., and Wall, C. (1998). UK Construction Skills in the Context of European Developments. *Construction Management and Economics*, 16(5), 553–567.

Coffield, F., Moseley, D., Hall, E., and Ecclestone, K. (2004). *Learning Styles and Pedagogy in Post-16 Learning: A Systematic and Critical Review*. London: Learning and Skills Research Centre.

Egan, J. (1998). *The Egan Report – Rethinking Construction, Report of the Construction Industry Taskforce to the Deputy Prime Minister*. London: HSMO.

Elmualim, A., and Gilder, J. (2014). BIM: Innovation in Design Management, Influence and Challenges of Implementation. *Architectural Engineering and Design Management*, 10(3–4), 183–199.

Fischer, M. (2000). 4*D CAD-3D Models Incorporated with Time Schedule, CIFE Centre for Integrated Facility Engineering in Finland, VTT-TEKES, CIFE Technical Report*. Helsinki, Finland.

Fischer, M., and Kunz, J. (2004). The Scope and Role of Information Technology in Construction *CIFE Technical Report* (pp. 19). San Francisco: Center for Integrated Facility Engineering: Stanford University.

Friedman, T.L. (2005). *The World Is Flat: A Brief History of the 21st Century*. New York: Farrar, Straus and Giroux.

Fruchter, R. (1998). Internet-Based Web Mediated Collaborative Design and Learning Environment, in Smith, I. (ed.), *Artificial Intelligence in Structural Engineering Lecture Notes in Artificial Intelligence* (pp. 133–145). Berlin: Heidelberg: Springer-Verlag.

Fruchter, R. (2004). Degrees of Engagement in Interactive Workspaces. *International Journal of AI & Society*, 19(1), 8–21. doi: 10.1007/s00146–004–0298-x.

Garris, R., Ahlers, R., and Driskell, J.E. (2002). Games, Motivation, and Learning: A Research and Practice Model. *Simulation Gaming*, 33(4), 441–467.

Gelb, M.J. (1998). *How to Think Like Leonardo da Vinci*. New York: Random House.

Gelb, M.J. (1999). *How to Think Like Leonardo da Vinci, Workbook*. New York: Random House.

Gelb, M.J. (2004). *Da Vinci Decoded: Discovering the Spiritual Secrets of Leonardo's Seven Principles*. New York: Bantam Dell.

Goulding, J., Nadim, W., Petridis, P. and Alshawi, M. (2012). Construction Industry Offsite Production: A Virtual Reality Interactive Training Environment Prototype. *Advanced Engineering Informatics*, 26(1), January 2012, 103–116, http://dx.doi.org/10.1016/j.aei.2011.09.004.

Goulding, J.S., and Rahimian, F.P. (2012). Industry Preparedness: Advanced Learning Paradigms for Exploitation. In A. Akintoye, J.S. Goulding & G. Zawdie (Eds.), *Construction Innovation and Process Improvement* (pp. 409–433). Oxford, UK: Wiley-Blackwell.

Goulding, J.S., Rahimian, F.P., and Wang, X. (2014). Virtual Reality-Based Cloud BIM Platform for Integrated AEC Projects. *Journal of Information Technology in Construction (ITCON)*, 19(Special Issue BIM Cloud-Based Technology in the AEC Sector: Present Status and Future Trends), 308–325.

Goulding, J.S., and Syed-Khuzzan, S. (2014). A Study on the Validity of a Four-Variant Diagnostic Learning Styles Questionnaire Framework. *Journal of Education + Training*, 56(2–3), 141–164, doi: 10.1108/ET-11–2012–0109.

Hou, L., Wang, X., Bernold, L., and Love, P.E. (2013). Using Animated Augmented Reality to Cognitively Guide Assembly. *Journal of Computing in Civil Engineering*, 27(5), 439–451.

Hou, Y., and Ji, L. (2010). Stimulating Design Creativity by Public Places in Academic Buildings. *Journal Structure and Environment*, 3(2), 5–13.

Johansson, F. (2004). *The Medici Effect*. Boston, MA: Harvard Business School Press.

Juárez-Ramírez, R., Sandoval, G.L., Cabrera González, C., and Inzunza-Soberanes, S. (2009). Educational strategy based on IT and the collaboration between academy and industry for software engineering education and training. *m-ICTE 2009, V International Conference on Multimedia and ICT's in Education*, Lisbon, Portugal, 172–176.

Khanzade, A., Fisher, M., and Reed, D. (2007). *Challenges and Benefits of Implementing Virtual Design and Construction Technologies for Coordination of Mechanical,*

Electrical, and Plumbing Systems on Large Healthcare Project. CIB 24th W78 Conference, Maribor, Slovenia, 205–212.

Kim, B., and Chris, S. (2001). Accommodating Diverse Learning Style in the Design and Delivery of On-Line Learning Experiences. *International Journal of Engineering Education*, 17, 93–98.

Klimecki R., and Lassleben H. (1998). Modes of Organisational Learning: Implications from an Empirical Study. *Management Learning*, 29(4), 405–430.

Kolb, D.A. (1984). *Experiential Learning: Experience as the Source of Learning and Development.* Upper Saddle River, NJ: Prentice-Hall Inc.

Kuhl, F., Weatherly, R., and Dahmann, J. (2000). *Creating Computer Simulation Systems: An Introduction to the High Level Architecture.* Upper Saddle River, NJ: Prentice-Hall PTR.

Kumaraswamy, M. (1997). Improving Industry Performance through Integrated Training Programmes. *Journal of Professional Issues in Engineering Education and Practice*, 123 (3), 93–97.

Laird, D. (2003). *New Perspectives in Organisational Learning, Performance, and Change: Approaches to Training and Development* (Third edition): New York: Perseus Books Group.

Latham, M. (1994). *Constructing the Team, Joint Review of the Procurement and Contractual Arrangements in the UK Construction Industry, Final Report.* London: HSMO.

McAdam, R., Leitch, C., and Harrison, R. (1998). The Links between Organizational Learning and Total Quality: A Critical Review. *Journal of European Industrial Training*, 22 (2), 47–56.

Mole, T. (2003). *Mind Your Manners: Business Cultures in Europe – Managing Business Cultures.* London: Nicolas Brealey Publishing.

Naoum, S., and Hackman, S. (1996). Do Site Managers and the Head Office Perceive Productivity Factors Differently? *Journal of Engineering, Construction and Architectural Management*, 3(12), 147–160.

O'Brien, L. (1989). Learning Styles: Make the Student Aware, *NASSP Bulletin*, 73(519), 85–89.

Oxford, R.L., and Ehrman, E. (1993). Second Language Research on Individual Differences. *Annual Review of Applied Linguistics*, 13, 188–205.

Pemberton J.D., and Stonehouse G.H. (2000). Organisational Learning and Knowledge Assets – an Essential Partnership. *The Learning Organization*, 7(4), 184–193.

Petric, J., Maver, T., Conti, G., and Ucelli, G. (2002). Virtual reality in the service of user participation in architecture. *CIB W78 Conference*, 12–14 June, Aarhus School of Architecture.

Pour Rahimian, F., Ibrahim, R., Rahmat, R.W.B.O.K., Abdullah, M.T.B., and Jaafar, M. S.B.H. (2011). Mediating Cognitive Transformation with VR 3D Sketching During Conceptual Architectural Design Process. *Archnet-IJAR, International Journal of Architectural Research*, 5(1), 99–113.

Rahimian, F.P., and Ibrahim, R. (2011). Impacts of VR 3D Sketching on Novice Designers' Spatial Cognition in Collaborative Conceptual Architectural Design. *Design Studies*, 32(3), 255–291.

Regenbrecht, H., and Donath, D. (1997). Architectural Education and Virtual Reality Aided Design (VRAD), in Bertol, D. (ed.), *Designing Digital Space – An Architect's Guide to Virtual Reality* (pp. 155–176). New York: John Wiley & Sons.

Salama, A.M. (2008). A Theory for Integrating Knowledge in Architectural Design Education. *Archnet-IJAR, International Journal of Architectural Research*, 2(1), 100–128.

Sampaio, A.Z., Ferreira, M.M., Rosário, D.P., and Martins, O.P. (2010). 3D and VR Models in Civil Engineering Education: Construction, Rehabilitation and Maintenance. *Automation in Construction*, 19(7), 819–828.

Sampaio, A.Z. and Henriques, P.G. (2008). Visual simulation of previous term civil engineering next term activities: didactic virtual previous term models. *WSCG 2008, 16th International Conference in Central Europe on Computer Graphics, Visualization and Computer Vision, Plzen, Czech Republic*, 143–149.

Santos, E.T. (2009). Building information modeling and interoperability. *XIII Congress of the Iberoamerican Society of Digital Graphics – From Modern to Digital: The Challenges of a Transition Sao Paulo*, 16–18 November, 2009, Brazil.

Schueller, S.M. (2012). Positive Psychology, in Ramachandran, V.S. (ed.), *Encyclopedia of Human Behavior* (Second edition) (pp. 140–147). San Diego: Academic Press.

Sternberg, R.J. (1985). *Beyond IQ: A Triarchic Theory of Intelligence*. Cambridge: Cambridge University Press.

Sternberg, R.J. (1996). *Successful Intelligence*. New York: Simon & Shuster.

Sternberg, R.J. (1997). A Triarchic View of Giftedness: Theory and Practice, in Coleangelo, N. N and Davis, G.A. (eds.), *Handbook of Gifted Education* (pp. 43–53). Boston, MA: Allyn and Bacon.

Succar, B. (2009). Building Information Modelling Framework: A Research and Delivery Foundation for Industry Stakeholders. *Automation in Construction*, 18(3), 357–375.

Succar, B. (2011). Organizational BIM: How to assess and improve your organization's BIM performance. *Revit Technology Conference Australasia*, V1.1, session 3 part B.

Thai, A.M., Lowenstein, D., Ching, D., and Rejeski, D. (2009). Game Changer: Investing in Digital Play to Advance Children's Learning and Health: The Joan Ganz Cooney Center. www.healthgamesresearch.org/our-publications/research-briefs/Game-Changer. [Accessed Date 22nd July 2011].

Uygun, Ö., Öztemel, E., and Kubat, C. (2009). Scenario Based Distributed Manufacturing Simulation using HLA Technologies. *Information Sciences*, 179(10), 1533–1541. doi: http://dx.doi.org/10.1016/j.ins.2008.10.019.

Vincent, A., and Ross, D. (2001). Personalise Training: Determine Learning Style, Personality Types and Multiple Intelligence. *The Learning Organisation*, 8, 36–43.

Wang, X., and Dunston, P.S. (2013). Tangible Mixed Reality for Remote Design Review: A Study Understanding User Perception and Acceptance. *Visualization in Engineering*, 1(1), 1–15.

Wang, X., Love, P.E., Kim, M.J., and Wang, W. (2014a). Mutual Awareness in Collaborative Design: An Augmented Reality Integrated Telepresence System. *Computers in Industry*, 65(2), 314–324.

Wang, X., Truijens, M., Hou, L., Wang, Y., and Zhou, Y. (2014b). Integrating Augmented Reality with Building Information Modeling: Onsite Construction Process Controlling for Liquefied Natural Gas Industry. *Automation in Construction*, 40, 96–105.

Watkins, K., and Marsick, V. (1992). Building the Learning Organization: A New Role for Human Resource Developers. *Studies in Continuing Education*, 14(2), 115–129.

Watson, J., and Hardaker, G. (2005). Steps Towards Personalised Learner Management System (LMS): SCORM Implementation. *Campus-Wide Information Systems*, 22, 56–70.

Wellings, J., and Levine, M.H. (2010). The Digital Promise: Transforming Learning with Innovative Uses of Technology: A White Paper on Literacy and Learning in a New Media Age. *Joan Ganz Cooney Center at Sesame Workshop*, www.digitalpromise.org/Files/Apple.pdf. [Accessed Date 22nd July 2011].

Whyte, J., Bouchlaghem, N., and Thorpe, A. (1998). *The promise and problems of implementing virtual reality in construction practice.* Paper presented at the Life-Cycle of Construction IT Innovations: Technology Transfer from Research to Practice (CIB W78), Stockholm, 3–5 June, 1998.

Wolstenholme, A. (2009). *Never Waste a Good Crisis: A Review of Progress since Rethinking Construction and Thoughts for Our Future.* London: Constructing Excellence.

Yoh, M. (2001). *The Reality of Virtual Reality.* Paper presented at the Seventh International Conference on Virtual Systems and Multimedia (VSMM'01), Organized by Center for Design Visualization, Berkley, CA: University of California Berkley, 25–27 October, 2001.

Youmans, R., and Arciszewski, T. (2014). Design Fixation: Classifications and Modern Methods of Prevention. *Artificial Intelligence for Engineering Design, Analysis and Manufacturing*, 28(02), 129–137

Zhang, X., Wang, H., Ma, H., and Wang, H. (2012). The Research of Digital Proving Ground Simulation System Based on HLA. *Procedia Engineering*, 29(0), 3624–3630. doi: http://dx.doi.org/10.1016/j.proeng.2012.01.542.

Zheng, X., Sun, G., and Wang, S.W. (2006). An Approach of Virtual Prototyping Modeling in Collaborative Product Design, in Shen, E.A. (ed.), *International Conference on Computer Supported Cooperative Work in Design* (pp. 493–503). CSCW 2005, LNCS 3865.

Zudilova-Seinstra, E., Adriaansen, T., and van Liere, R. (2009). *Trends in Interactive Visualization: State-of-the-Art Survey.* London Springer-Verlag.

10 Conclusion

Online learning for STEM subjects

Mark Childs and Robby Soetanto

Understanding the student perspective

Throughout the various cases described in this book, a common feature is the wealth of data produced when students are asked to supply their experiences of learning. When presenting these data, the findings are often presented in different ways for different audiences throughout the dissemination phase of a project. One approach to presenting our findings that generated particular interest amongst audiences was the collection of *Things students learnt from the online synchronous collaboration that were basic online communication skills.* We had, actually, made the erroneous assumption that they were self-evident. In summary, these were:

- Not breaking off for private conversations. In some cases, students report the participants at other sites muting their microphones in order to have private conversations. This would not be acceptable in a face-to-face meeting and so seems very odd that students would do it in a videoconferencing situation. One personal reflection by a student stated that "In some of the early meetings we had, there was some conflict . . . over how to proceed with certain points particularly with the information we were given. This did cause some tension particularly when microphones were muted and private discussions could be seen on the screen. Some students however felt that this was acceptable behaviour and saw it as a benefit of using videoconferencing. However, this act did create a division between team members who were collaborating online and so had a negative impact on the level of trust."
- Effectively supplementing face-to-face behaviours to compensate for less physical presence. Videoconferencing is a more limited form of interaction than face-to-face situations, exacerbated by the Internet connections in some cases not being effective. Students struggled with adapting to this in several instances, particularly when questioning each other. Techniques such as alerting a person to a one-to-one communication using their name, rather than directing one's gaze, had to be learnt. One student reflection was that videoconferencing "was a new experience, communicating through a webcam is very different to face-to-face interaction and I found it harder to translate

ideas and more of a struggle to build a rapport with teammates. I think the reason for this, is the fact we are communicating through a webcam and not face to face, so body language was harder to interpret and grasping each individual's understanding was more of a struggle." The students found some way to compensate for the limitations. "To prevent having to repeat oneself, speaking slowly and clearly was vital. Additionally, because we were talking to a screen, directing conversation was problematic. I found the best way to deal with this was to state each individual's name when addressing them."

- Formally structuring meetings so that only one person speaks at once, and all people get a chance to speak. This does not appear to have been learnt in all cases. As one student noted, "I don't think we were able to achieve effective collaboration through meetings in the final stages, because there was less structure and people were talking over one another. I found a good way to deal with this is to request the chance to speak, before offering an opinion, this way you have full attention of the group."

- Planning and structuring meetings. Students struggled through some meetings because they were used as a means to transfer information rather than discuss concepts. Circulating agendas before a meeting was still not common practice with all the groups by the end of the first semester. When reflecting, a typical observation by students was, "I think if we had used an agenda as frequently as in stage 2, it would have helped structure the meetings." Determining action points arising from meetings was also a practice that emerged over time rather than being in place from the outset, with comments from students being made such as, "Some delays weren't because of the lack of dedication, just simply circumstances where confusion has arisen over some tasks."

- Ensuring everyone is included. Some groups reported lapses in inclusivity, caused by the tendency for some people to dominate the videoconference and not employ techniques to include those who were finding it difficult to engage, even ones as simple as pausing occasionally and checking if anyone wanted to ask questions. As one student noted, "we could have asked if they were ok on a regular basis, just to acknowledge their presence, and to allow them to ask questions if needed. . . . They were quiet people and so checking if they are ok is a good way to feel welcome."

- The students seemed prepared to put up with the echo rather than insisting on it being stopped. Some students seemed unaware that a simple remedy such as using a headset would prevent the echo from occurring, perceiving it as a problem with the quality of their hardware, for example: "Sometimes I could hear my own voice after I spoke, which caused me the biggest problem. I don't see this as a problem we need to worry about, since in the industry, companies have the very best in Internet and software packages, to prevent this occurring."

- Most troubling was that they did not take into account peers with disabilities. One student amongst the participants was deaf, and despite this, the team still used sound for their meetings. This meant that he could not participate in the meetings, as the webcam quality was also too limited for him to lip-read unless he was addressed directly, in which case his colleagues at his end could interpret for him – a task they could not do the rest of the time, as their attention was taken up with participating in the meeting. This student concluded: "This is something that can be learnt from by society in terms of working in teams with different people, where there may be language barriers or, in my case, hearing difficulties. In order to overcome this situation, the team needs to discuss these problems and devise ways of alternative communication." None of the students seemed aware that this problem could be avoided simply by switching to using text chat in the meetings.

- Working in unsuitable areas. Students reported problems with connectivity due to working in places with poor Wi-Fi and background noise making communication difficult. One student concluded that "Locating self in more silent zone while being on meeting will help to eliminate the noise disturbance issues and communication might be more efficient."

The failure to respond to issues with the hardware and find workarounds or establish best practice is in contrast to the flexibility students showed when confronting issues about choice of platform; when responses did not happen with one medium, such as email, students set up alternative discussions in Facebook, for example, or selecting a common platform for sharing documents. These underline three aspects of working with students in an online mode that, despite the common assumptions about so-called Generation Y or Generation Z students (key learning points are numbered):

1   Many students have limited experience in the use of synchronous communication software.
2   Experience with the use of collaborative technologies such as social media does not translate to the ability to use this in learning and working scenarios.
3   Students often display low self-efficacy when adapting to limitations of hardware, although higher self-efficacy when choosing appropriate software.

Online collaboration requires the students to learn a new set of skills in addition to learning the subjects. Although it may be burdensome for some, it can bring positive outcomes which can benefit their own employability following graduation and beyond. In this case, Soetanto *et al.* (2014) found that participating students have become better prepared in the management of collaborative design projects. This is also suggested within Soetanto and colleagues' chapter, referring to 'proactive personality' coined by Tymon (2013). A key question that remains problematic is to what extent do you train students in the use of technology and in how to conduct online collaborations, and to what extent should they be left to ascertain these for themselves? Although learning by experience becomes more deeply embedded and so should be encouraged, there are some aspects which should be scaffolded. Identifying which is perhaps a personal choice of the lecturers. However, we would recommend:

4   Making students aware of their responsibilities regarding equality of access, across differences in culture, experience and ability.
5   Supporting students to develop the critical reflection skills to enable them to learn effectively from their experiences.
6   The tutors to instil and nurture confidence and self-efficacy, and communicate the benefits continuously from the start of implementation.

There are also a sizeable minority of students who struggle with experiencing online interactions as "authentic" relationships. Given that an essential part of collaborative learning is learning *together*, and that social exchange underpins the trust that helps create effective relationships, then this difficulty can be a severe limitation to the formation of effective collaborations. Indeed, online collaborative learning per se can often prove problematic, with many students not having experienced the basic demands of conducting formalised project management during their degree courses, as the close proximity of peers enables group work to be successfully completed through *ad hoc* daily encounters. Simple practices such as frequently checking emails and arranging meetings and setting action points were not developed throughout their previous experiences of collaboration.

Furthermore, the role of socialisation online and the ability to project and perceive social presence is a factor in creating online rapport, which supports the cycle of building and maintaining trust, and yet despite social media being a familiar context in which students operate (all of the students questioned, for example, used Facebook in their social life), in the lead case study of this book, few socialised with their peers in other institutions online or formed strong social bonds. This may have been reticence about appearing unprofessional, indicating a misperception about what constitutes professionalism, but did have a deleterious effect on the building of a co-operative working relationship. These factors in tandem lead to the following points:

7   Many participants feel anxiety about purely online interactions and struggle to develop the same rapport in them as in face-to-face ones.
8   The move from offline to online collaboration is not only demanding due to the online nature of the interaction but also due to this being the first time when *ad hoc* interactions do not suffice, and so many basic skills associated with face-to-face (but formal) collaborations also need to be learnt.
9   Students need to be aware of the value of forming social connections in online collaboration and be prepared to dedicate time to achieving these.

As can be seen in the two chapters by Childs, the number of students who cannot be relied upon to contribute professionally to a collaboration is a minority but is large enough that increasing the interaction from four to six participants noticeably increases the chances of introducing a non-performing member to a consortium. "Carrying" a non-performing member of a group continues to be a source of anxiety for students, and the role of peer assessment is seen predominantly as existing as a recourse to "punish" non-performing members of a group.

The value and requirements of online collaborative learning activities

This does raise the questions that, if it is so contentious and problematic, ought collaborative practice be introduced to the curriculum? As noted in Chapter 1 by Soetanto and colleagues, there can be numerous nervous responses to any development that might have a negative impact on student satisfaction due to relatively uncertain outcomes. However, Soetanto and colleagues in Chapter 4 discovered that the employability attributes were perceived by the students as being improved following participation in the collaborative design project. Further, the evidence for the effectiveness of collaboration in online learning can be seen in the chapter by Smith and colleagues, the social connections that come with interaction encourage learning; in their comparison of two MOOCs, the one that supported more interaction and collaboration was more successful with the students.

Furthermore, there is a need for students in STEM subjects to acquire the skills for those industries in which as graduates they are likely to find work. The roles they are most likely to take as STEM graduates will require the ability to work in distributed teams, and so when looking at "soft skills" to include in the curriculum, this becomes essential. In our experience there is no greater proponent of this development than the students themselves. When asked, the students of the Coventry, Ryerson and Loughborough case study indicated these reasons taking part was of value to them. These were:

- Working in international teams
- Working in larger groups
- Working in multidisciplinary teams
- Working with people from outside the institutions and therefore having to present a "professional" persona
- Working in different roles (in those collaborations which enabled this)
- CV enhancement
- Greater authenticity of the exercise
- To stretch themselves. As one student stated, "I chose to be part of this task in order to be able to step out of my comfort zone. At university, I have worked with many students on numerous projects and it has worked well and we had no major difficulties. I feel I was getting too comfortable and that I wasn't putting myself out there. I therefore decided to sign up to this project because I strongly believed that it would be a positive experience."

These perceptions by the students of the value of the exercise are confirmed by studies of the increase in self-efficacy, as noted by Soetanto and colleagues in

Chapter 4. This held true even for those students who had not had successful collaborations. All felt it had been a worthwhile process, even though, or rather because, it had been a difficult experience; the difficulties were perceived as being an authentic preparation for what they were likely to experience within their professional practice.

Making online collaboration more effective

Given that collaboration can be difficult but is recognised as worthwhile, the proper course would therefore evidently be to introduce online collaboration but to introduce practices that ameliorate as much as possible the external factors that have a negative impact on the students' experience. In the model given in Chapter 3, the tutors can improve as much as possible the design and support of the learning activities.

Learning design is evidently key to all effective teaching. The creation of tasks that reflect course content, provide authentic learning experiences and are engaging is an intrinsic part of the skill set of any lecturer. However,

10 Developing online collaborative tasks between institutions creates an additional series of requirements for learning design and assessment that even the most experienced of teachers may be unprepared for.

The first of these is the very different demands of the separate institutions involved. Differences in assessment, particularly different approaches to peer assessment, the timing of examinations and particularly the scheduling of teaching weeks all have an impact on the effectiveness of the collaborations. The impact of these can be lessened by designing the activity so that, in the case of the different schedules, students at different universities have different tasks to be completed at different times. This can more accurately reflect the situation in the workplace, where distributed teams have different roles, but it can give rise to another issue perceived by students as a significant one, that of role allocation within the collaborations. Starting first can be perceived as giving an advantage in terms of leading the collaboration and so, as one student stated, "put them in a position of power as they had more of an understanding of the brief. This is slightly unfair because it meant that other universities were less able to contribute towards the early stages of the project." These concerns are exacerbated by students seeing the task as an opportunity to develop experience and their awareness of the fact they will be assessed on their work adding stressors that would not occur in the actual work-based scenario the task is aiming to emulate. Assigning roles based solely on discipline, for example, project leadership taken on by those studying project management, is resisted for similar reasons. As one student stated, "(we) were both forewarned by our project tutor that we should be prepared to tussle over the project roles early on with the other students, as there was always a chance that the students from the different universities would be 'stuck' with work specifically for their study discipline. . . . This was a slight and I felt it hampered effective relationships early on due to the lack of acknowledgement of mine and other colleagues' experience . . . potentially leaving other members feeling a lack of worth."

To an extent, the fragmentation into separate student roles is driven by disciplinary difference but also because, despite the authenticity of the tasks set, the unavoidability of this being set within the context of degree courses can lead to students being strategic about their efforts. The time overhead in communicating and collaborating may tempt students to subdivide the tasks in order that they can be conducted individually rather than properly collaborate, as described by one student who stated that "we immediately split up the work rather than attempting to discuss the objectives and the best way to achieve them. . . . These were then completed in isolation, produced into one document and submitted. There was no agreement or group consensus, it was merely an exercise to gain marks for that section. In practice this meant we had a fragmented plan that no individual had bought into."

Although the degree to which institutions can adapt to differing schedules is limited, where planning can make the most difference is in co-ordinating the brief given to students and in making the assessment process as equitable as possible. Often validation processes and course requirements can lead to demands on changes being made, often with insufficient flexibility or time to liaise with partners in other institutions, so the task outline and the amount of marks the separate parts of the assignment are worth can end up slightly different. As long as the students are all told the same thing about what they have to do for their group task and are aware of the marking criteria their peers at other institutions are subject to, then differences are acceptable. It is transparency that is key. For this reason:

11   Create a common portal for all of the participants at all of the participating universities, containing all key information, so that students can also see what their peers at other institutions are being told.

In our experience, this was the one aspect of support of the student experience that would be the simplest to improve.

Even if the value of online collaboration as an activity is accepted, this may not necessarily mean that this requires inter-institutional collaboration given the issues of aligning schedules, brief and assessment criteria. We would argue, however, that

12   Inter-institutional collaborations are far more effective learning scenarios than intra-institutional collaborations.

The reasons for this are that students, in our experience, presented a far more committed and formal approach to the collaboration when they felt that they were representatives of their university to other universities. A more compelling observation was the contrasting experiences of the undergraduates and postgraduates recounted in Chapter 3; when the online aspect wasn't essential, that is students were required to interact online solely to give them experience of interacting online rather than because there was no other way to collaborate, it was predominantly rejected as having no value. To produce an authentic experience of online collaboration that has meaning, the other members of the group have to be distant. Other differences provided by international working are also valuable; through this students can learn to incorporate other legal requirements, other terminologies (even terms such as "first floor" vary from country to country) and also a surprisingly complex skill which students struggled to master, working across time zones. Thus, the experiences of working cross-culturally are also of value. Here this is not so much the culture of working with other nationalities (as most institutions will have multinational students) but the culture of other disciplines and institutions is a valuable learning experience. Nevertheless, as found and described in Chapter 4, a successful collaborative project would require:

13 Participating students to be equipped with and/or possess sound technical knowledge, applied using good teamwork skills, understanding direct and wider implications of design decisions and actions with the team.

Development of practice in online collaboration

As can be seen throughout the book, the authors have adopted a range of models to help understand the student experience. The main case study consists of two iterations of a similar process, and the development across the two and analysis of the differences was guided by action research, the separate cycles described by Childs in his two chapters and the process by which this action research informed this process explored by Tolley and Mackenzie in their chapter. The models developed to segment the student experiences and make them more explicable themselves were developed across these iterations, the first model being an input–process–output model (used by both Childs and Soetanto and by Rahimian *et al* in their respective chapters), the second a pyramidal model in which each aspect of experience builds on the level below and articulates the experience of staff and students in more detail. Tolley and Mackenzie also draw on Winnicott's principles of academic play, Soetanto and colleagues on Bandura's theories of self-efficacy and Rahimian et al on the triad of successful intelligence, positive psychology and appreciative intelligence. Although disparate in origin, all of these approaches have in common the focus on the development of a learner's sense of self and identity, in which they are given the locus of control over their own learning and, further, the awareness that they *have* control over their own learning. This meta-cognitive approach is fundamental to any process in which students are handed the reins of developing their practice. If learners are being provided with experiences from which they can learn, a mechanism at the centre of learning by design, then this learning can be successfully acquired and embedded by providing a framework for students to reflect upon the challenges and solutions they face.

These challenges faced by students in online collaboration are various. The difficulties some face in establishing rapport in online interactions is one. The translation of the normal social bonds in face-to-face interactions with peers to those with peers in an online collaborative situation is another. Incompatibilities with software and limitations of the hardware also limit the ease of interactions. Responding to those situations when collaborators underperform and identifying ways to rebuild trust is possibly the greatest. Conducting these in an environment where results are graded and where they may have to assess peers adds an additional level of anxiety.

Where we as educators may be prone to not support our students as effectively as we could is in overestimating the complexity of these issues not only for our students but also for ourselves; hence the following observation:

14 We also need to be aware of our own limitations in online collaboration and ensure that we put into practice the skills that we expect our students to acquire.

Inter-institutional online collaborative learning activities have as a prerequisite (evidently) inter-institutional online collaboration between teachers at different institutions, and this can be a stumbling block, especially when the tendency to focus on our immediate teaching, our own institution's concerns and our own course administrative needs can seem paramount. This is why the meta-evaluative practice espoused by Tolley and Mackenzie should lie at the centre of how we enact practice and learn from it.

Whether these issues continue to dominate online collaborative practice is another question. The growth of MOOCs as described in Smith and colleagues' chapter is widening the audience for online collaborations; the developments in cloud computing as described by Li and Migliaccio and colleagues has far-reaching implications for how ubiquitous computing can change practice in education. The cases also of virtual reality, as described by Migliaccio and colleagues and Rahimian and colleagues may have the greatest impact on collaborative practice. The issues of forming presence and copresence, which lie at the heart of learning *together*, will be transformed by the introduction of virtual reality and augmented reality. As stated in the introduction, the learners we will soon be encountering within our higher education institutions, the so-called Generation Z, will have been born in the 21st century, and although they still struggle with expressing themselves online, these are technologies they will have grown up with. The children being born now are still without a singly agreed upon reductionist label, but we would perhaps categorise them as "Generation &." Generation & will experience the material world and virtual reality and augmented reality as a seamless whole. The sense of switching between offline and online modes and having to establish new skills to create a sense of togetherness online may disappear as the technology discussed in the Rahimian and colleagues chapter, develops. Identifying the skills now, developing students' and our practice to anticipate employment needs and changes in technology and identifying the best of the pedagogical theories to inform that practice, will ensure the best possible futures for every generation of students.

References

Soetanto, R., Childs, M., Poh, P., Austin, S., and Hao, J. (2014). Virtual Collaborative Learning for Building Design: Proceedings of the Institution of Civil Engineers – Management. *Procurement and Law*, 167, MP1, 25–34. http://dx.doi.org/10.1680/mpal.13.00002.

Tymon, A. (2013). The Student Perspective on Employability. *Studies in Higher Education*, 38(6), 841–856.

Index